SEARCHING FOR COMETS

SEARCHING

FOR COMETS

Deciphering the Secrets of our Cosmic Past

Louis Brewer Hall

McGRAW-HILL PUBLISHING COMPANY

New York St. Louis San Francisco Bogotá
Hamburg Madrid Mexico Milan
Montreal Paris São Paulo Tokyo Toronto

1 2 3 4 5 6 7 8 9 DOC DOC 8 9 2 1 0 9

ISBN 0-07-025633-0

Library of Congress Cataloging-in-Publication Data

Hall, Louis Brewer.
Searching for comets : deciphering the secrets of our cosmic past / by Louis Brewer Hall.
p. cm.
ISBN 0-07-025633-0
1. Comets. I. Title.
QB721.H25 1989 89-14525
523.6—dc20 CIP

Book design by Eve L. Kirch

Dedicated to our family of pioneers:

Francis R. Scobee
Michael J. Smith
Judith A. Resnik
Ellison S. Onizuka
Ronald E. McNair
Gregory B. Jarvis
S. Christa McAuliffe

CONTENTS

PREFACE

This book developed from a program that was generously sponsored by the National Science Foundation and designed to interest students without a science background in Comet Halley and the Comet Halley spacecraft missions. Professors trained in the humanities worked with established scientists and engineers. I owe a great debt of gratitude to:

Dr. Edgar Everhart, director of the Chamberlain and Mt. Dick observatories, one of the leading authorities on comet orbits, illustrated the basics of astronomy and unraveled the intricacies of comets with a rare combination of simplicity and excitement.

Dr. Clyde Zaidins, professor of physics at the University of Colorado at Denver, helped me immensely with the beauties of spectroscopy.

Dr. Nelson Fuson, professor emeritus of physics at Fisk University and director of the Infrared Institute, carefully perused the manuscript for errors in the dynamics of space.

To Dr. Hans Liebe, of the National Bureau of Standards, I owe heartfelt thanks for great help with Giotto.

The scientists and engineers of the Laboratory for Atmospheric and Space Physics (LASP) at the University of Colorado in Boulder made their facilities a second home to me with their boundless hospitality and cooperation. Their pride in Spartan-Halley and the expertise necessary to design and construct it was contagious, and I hope I have reflected something of their great, though disappointing, achievement.

I can mention only a few by name: Charles Barth and Fred Wilshusen, Sam Jones, Bob Leyner, Neil White, Rick Kohnert, Ross Jacobsen, Alan Stern, Paul Bay, Beth True, Yvonne West.

The engineers and scientists of the Goddard Space Flight Center were partners of LASP in constructing and integrating the Spartan-Halley spacecraft, and I want to thank especially Morgan Windsor, the Spartan-Halley project manager, for his insights and help.

I was fortunate in having Alan Delamere and Harold Reitsema not far away at the Ball Aerospace Systems Division. They guided me through the intricacies of one of the great engineering achievements of the Halley spacecraft, the multicolor camera on Giotto. Takehiko Kobayashi and Michio Hirakawa helped me understand more clearly some of the complications of Suisei and Sakigake. I hope what I have written is worthy of their patient and painstaking assistance. They are all great teachers, and I hope the pages of this book have preserved their dedication and joy in what they have accomplished.

Learning about comets and the engineering of spacecraft has been a rewarding experience, one full of excitement and challenge. I have tried to follow my informed and expert sources with greatest care. Undoubtedly, I have made errors. They are mine alone and in no way reflect on the expertise and knowledge of those who have so graciously helped me.

1

Fireballs to Snowballs

The fiery passage of comets across the skies has been a source of wonder and mystery throughout human history. A comet appearing after sunset or before sunrise can be the brightest object in the sky, as bright as the Moon, so bright it blots out nearby stars. Comets intrigued ancient scientists who could observe with only their naked eyes. Scientists in Babylon, Egypt, China, and Greece all wondered about comets. Why did they suddenly appear and as suddenly vanish? What were they? Why were they there?

During the fourth century B.C. Aristotle devised a cometary theory without performing experiments, nor did he seem to have observed comets, or even the planets, very assiduously. He worked deductively, that is he started with a general concept and then applied it to specific phenomena. The outermost part of the Earth, he said, was dry and warm, and under certain damp conditions, exhalations from the Earth arose into a circle of pure fire—invisible because it was pure—lying between the

Earth and the Moon. If these exhalations were the right consistency, they ignited and became balls of fire, leaving behind them fringes, which reminded people of flowing hair. The Greeks, and after them the Romans, called these objects "hairy," *cometa*. Because he did not use instruments of any kind, Aristotle could not measure the movement of comets across the sky nor measure their distance from the Earth, which he assumed was a stationary platform. His theory was accepted in the West for 1800 years.

Five hundred years later, in the third century A.D., across the Earth from Athens, in Loyang, China, Chang Heng, grand lord chronologer of the emperor of China's astronomical bureaucracy, was also observing and recording the stars, planets, and comets. He was using instruments that he had designed and constructed himself. For timing he built a very accurate water clock, and to locate the stars he built an armillary sphere. This consisted of intersecting brass rings, which represented the paths of the Sun, Moon, and some of the constellations moving horizontally around a fixed point on the Earth. We are not sure how, but somehow his armillary was mechanized by water power, and he used this instrument to keep records not only of the regular movements of the heavens, but also irregular occurrences like the arrival and departure of comets. Chang's own records have not survived directly, but Chinese records that made use of his have survived. In these is found the first recorded appearance of Comet Halley, May 239 B.C.

In the West, Aristotle's exhalation theory continued to be accepted, and it was not until the fifteenth century that astronomers had the instruments to be able to fix the position of comets in the sky with any degree of accuracy. George Purbach, an Austrian astronomer, and Paolo Toscanelli, an Italian, observed the Comet of 1456 move across the sky. By following its path not along the horizon, as had always been done previously, but along the ecliptic, the imaginary path of the Sun around the Earth, they measured how far the comet had moved each

24-hour period. However, neither Purbach or Toscanelli were able to calculate how far the Comet of 1456 was above the Earth.

Aristotle's theory that comets appeared between the Earth and the Moon wasn't finally disproved until over 100 years later when the comet of 1577 dazzled the imagination of European astronomers. One of these was Tycho Brahe, a Dane, and the most accurate observational astronomer who never held a telescope. His positioning of the stars, planets, and comets was so accurate that, even 100 years later, Edmond Halley and Isaac Newton could refer to his observations. Tycho observed through open sights—sights without lenses—that moved on the calibrations of a giant brass quadrant, 12 feet tall, 5 feet in radius, and fixed to the wall of his observatory. In addition he used a huge armillary, which 300 years previously had been introduced to Europe by the Arabs.

He calculated that the Comet of 1577 was not an atmospheric phenomenon at all. It was at least four times the distance between the Earth and the Moon. The comet, then, was truly a celestial object.

Tycho was a contemporary of Galileo's, and, of course, Galileo's use of the telescope is well known. In 1618–1619 three spectacular comets appeared in Europe and caused a debate on the composition of comets between Galileo and Horatio Grassi, the professor of mathematics at the Collegio Romano (now the Gregorian Institute) in Rome. At this time Galileo was too ill to use his telescope to observe those comets, but his friend Mario Guiducci observed the comet of 1619. About it Guiducci said, "I do not see that anything may be inferred about the comet from its slight enlargement by the telescope except that it is luminous." Galileo's telescope had the power of a large pair of binoculars today.

One hundred and twenty-eight years after Tycho and Galileo, Edmond Halley, with the help of his reclusive friend Sir Isaac Newton, demonstrated that comets are not just celestial vagrants but sail in ellipses. Halley calculated that the Comet of 1682 had passed the Earth and circled the Sun at least twice before, in 1531 and 1607, and that the same comet would appear again around Christmas in 1758.

It did, of course, and now Halley's name not only graces the comet whose ellipse he plotted, but is among the best known of astronomers. In the space of two generations in the seventeenth century, astronomers discarded myths and theories that had existed for 2000 years. They proved comets, like all celestial bodies, acted according to Newton's laws of motion. Since then, the return of Comet Halley has been predicted four times according to its 76-year ellipse, and each time the prediction has been more accurate.

Scientists in the eighteenth and nineteenth centuries who pondered the questions of the size and origin of comets found some answers. Lexell's Comet of 1770, for example, was proven to be much smaller than the Earth when its close passage to it produced no effect.

The technology needed to reveal the nature of comets began to be developed slowly through the nineteenth century. Sparked at first by the interest in optics that naturalists and gentlemen scholars pursued, their study of light produced important results.

We know that Newton discovered the individual components of light by placing a prism in front of a condensing lens, which divided the light into the colors of the rainbow. Then in 1802 the German optician, Joseph Fraunhofer, using a higher-powered prism, discovered hundreds of dark lines in the color spectrum of the Sun and other stars. Next, in the mid-nineteenth century, two other German scientists, Robert Bunsen and Gustav Kirchhoff, found that Fraunhofer's lines represented the absorption of light in the solar atmosphere by certain *identifiable* elements like sodium. Scientists realized that by using prisms to break light into a spectrum of colors they could determine the composition of objects millions of miles distant from Earth.

Astronomy was revolutionized. From this discovery the science of spectroscopy was born, and since then astronomy has been more precisely called astrophysics. Modern spectroscopic techniques are capable of determining the precise identification of each gas in a comet's atmosphere, as well as the temperatures, velocities, and particular chemistry of that atmosphere.

Spectrographs were first used to examine comets early this century. Beginning in 1908 with Comet Morehouse and then in 1910 with Comet Halley, astronomers like George Ellery Hale and Edward E. Barnard probed the composition of comets. In the spectrum of the comets' atmospheres they identified water, carbon dioxide, carbon monoxide, and among other elements, sulfur and cyanogen. They recorded the changing structures of the coma and tails of comets and made meticulous photographic records of their passage across the sky.

Meanwhile some of the most famous names in physics, Albert Einstein and Max Planck, developed an understanding of the behavior of matter on the largest scale, that of the cosmos, and on the smallest scale, that within the atom itself. Scientists developed new theories about the origins of the universe and the solar system and the evolution of comets.

According to one widely held hypothesis, a great cloud of dust and gas weighing anywhere from 1000 to 10,000 times our Sun began to collapse 4.6 billion years ago. The collapse continued as the mass within the contracting cloud exerted a gravitational force on the cloud itself, further increasing the rate of collapse, which in turn caused the cloud to heat up. A million years or so of this collapse resulted in the gas and dust at the center of the condensation reaching pressures and temperatures great enough to begin nuclear fusion, and the Sun ignited.

The debris left over from the Sun's formation was careening about, most of it in the form of a thick shell of gas and dust. This shell itself condensed. Near the Sun temperatures were so high that only metal-bearing molecules could condense, forming iron, feldspar, and a silicate of iron and magnesium. Further out where Mars would eventually form, water-ice condensed. Beyond that, in the further and colder reaches of the debris nebula, ammonia and methane formed. The solar system was gradually creating planets.

The evidence for this initial formation is largely circumstantial. However, studies of star formation across our galaxy provide many examples of this process. Recent observations made by the Infrared Astronomical Satellite (IRAS) point to collapsed dust and debris zones around a number of young stars, including our neighbor, Vega, only 27 light years away.

Following the Sun's ignition as a star and the initial collapse and condensation of the debris nebula, pebbles, rocks, and boulders began to form out of colliding dust grains, agglomerations of rock, and ice varying in size, objects known as planetesimals. These in turn were swept up in the gravitational fields of protoplanets forming among the chaos.

From these protoplanets there came first the inner planets: Mercury, Venus, Earth, and Mars. Jupiter and Saturn followed later, within perhaps 100 million years. Neptune and Uranus, by contrast, may have taken over a billion years to form in the slow-moving outer regions of the nebula.

Still, the myriad hunks of debris, the most numerous members of the solar system, remained. Much of this debris was swept up in collisions with the planets. However, there were many near misses as well. In some of these "close encounters" the debris was sent careening closer to the Sun where it evaporated in the intense heat. Still other particles were ejected from the planetary region altogether by the momentum of the revolving disk of the outer planet.

Modern astronomers, who have calculated the consequences of the primordial melee, have speculated that the orbits of the ejected debris particles range perhaps almost half-way to the nearest stars. Needless to say, such distant orbits take literally millions of years to complete. It is in these frigid and lonely orbits that between 100 and 200 *trillion* comets are believed now to reside.

They are frozen remnants of the debris nebula that evolved

into our Earth and the other planets. They can be called celestial Rosetta stones, archaeological artifacts that can help us decipher the mysteries of our cosmic past—and perhaps even our cosmic destiny.

In 1950 the Dutch astronomer Jan Oort conjectured that the solar system is surrounded by these frozen primeval remnants, actually a tremendous cloud of comets, an orbiting shell hundreds or thousands of times farther away than any of the planets. Left alone, the comets in this Oort Cloud would never be seen in the inner solar system.

The universe is, however, not quiet. It is filled with exploding stars, high intensity wavelengths, possibly black holes at the center of galaxies, and violent shock waves that blow across the void. Comets are so far away from the Sun that they can barely hold their tenuous orbits. Any push or shove could remove them from the Cloud altogether.

Oort demonstrated that the necessary pulls might come from passing stars and even from giant molecular clouds. Oort's statistical calculations showed that on hundreds of random occasions stars have actually swept close or even through the Cloud. Comets in the Cloud orbiting on faint gravitational strings can be either ejected from the solar system altogether by the gravitational field of a passing star or sent careening on paths that take the comets close to the Sun.

More recent researchers estimate that these close encounters with passing stars eons ago cast showers of millions of comets into the solar system, perhaps sometimes bombarding the planets, making the great craters that scar the Moon, Mercury, Mars, and the satellites of Jupiter and Saturn.

Astronomers now observe about a dozen comets a year that have made the several million year glide from the Oort cloud close enough to the solar system to be captured by Jupiter's gravitational field. They start their journey moving in the same direction as the other planets—counterclockwise as "seen" from the Sun or clockwise as Comet Halley moves.

As the comet continues on its journey and after a hundred

or more years crosses the orbit of Saturn, the Sun's feeble heat (at that point less than one-hundredth as intense as that which falls on the Earth) begins to evaporate the icy, age-old surface of the comet. Passing Saturn, each second the comet loses, perhaps, a gram or so of its surface ice by evaporation, about 1300 pounds of ice each hour of its passage. From this evaporation, a spherical cloud known as the coma forms around the solid body that is the comet itself.

As the comet passes Mars, and then the Earth, and approaches close to the Sun and then departs, those elongated tails keep streaming off the coma. On its closest approach to the Sun, its perihelion, the rate of evaporation can reach as high as several tons a second. Carried with this water in the effusive torrent are other vapors and dirt and dust that is embedded in the nucleus. The comet grows to several million miles in diameter, and the gases fluoresce in the strong ultraviolet sunlight of space, glowing as a rainbow of colors. Carbon monoxide and hydroxides each glow an eerie violet-blue, hydrogen and oxygen glow red, and other gases glow green and yellow. It was this bright florescence, invisible to us on Earth, that in 1973 so thrilled the astronauts in Skylab, the first human beings able to observe it.

It was the coma and tails of Comet Morehouse in 1908, of Comet Halley in 1910, and Comet Kohoutek in 1973, Comet Encke in 1974, Comet West in 1976, Comet Wild in 1980, and many others that astrophysicists analyzed with their spectroscopes and other instruments. They identified carbon molecules, hydrogen, hydroxide (one oxygen plus one hydrogen atom), cyanogen, and a host of other molecules in the coma.

In the 1950s Fred Whipple, at that time Director of the Harvard-Smithsonian Center for Astrophysics, constructed a detailed hypothetical model of a comet. In Whipple's paradigm, the comet itself is a mixture of water-ice, dust and dirt, and some traces of other components like dry ice (frozen carbon dioxide), methane, and ammonia. Whipple's model has been aptly called the "dirty snowball."

However, the real secrets of a comet are not in its coma and tails. Its real secrets are within the coma, in the heart of a comet—its nucleus, that relic of our cosmic past from the time of the solar system's birth. Comets are archaeological artifacts, 4.6 billion years old, eternally sailing through space.

To date, no Earth-bound instruments have been able to penetrate that coma of dust and gas surrounding the comet to observe the nucleus. But with both American and Soviet manned ventures into space, it became possible to contemplate observing comets at closer hand.

On July 12, 1972, American astronauts made their last Moon landing in Apollo 17. Whole Apollo modules and Saturn V rockets, their giant launch vehicle, had been stored unused. The third stage of the Saturn V was the S-IVB, an oxygen-hydrogen fuel tank, more than 20 feet in diameter and 28 feet high. Out of the S-IVB, the National Aeronautics and Space Administration created America's first space station, Skylab. Nine astronauts, in teams of three, successively conducted experiments, ate, slept, and exercised 270 miles above the Earth for 171 days between May 1973 and February 1974.

2

The Good Bird

When in the spring of 1973, NASA created out of a spare Saturn fuel tank a compact research park, it had a physics lab, biology lab, and an astronomical observatory from which the astronauts could observe the Earth, the Sun, stars, and a comet.

Living quarters were at one end of the rehabilitated S-IVB. Three private sleeping cubicles each contained a sleeping bag fixed to the wall. Floating off into dreamland is more than a metaphor in Skylab. Astronaut Ed Gibson slept head up, but after all, head up or down, sideways or backward, it didn't matter—orbit's weightless cushion provided posture-pedic rest unparalleled since our amphibian ancestors left the ocean.

Across the fuel tank was the dining room, known as the wardroom (Navy influence), with an 18-inch window. There was a small exercise area between the sleeping cubicles and the dining area. The bathroom had a real toilet with a seat, flushed by compressed air. The seat was against the wall, with handholds, footholds, and a seat belt. Partitions between the rooms were gray metal with oblong doorways that had no doors.

The ceiling was metal grillwork designed to allow specially cleated boots to lock in, providing an anchor for crew members to work.

There wasn't a true up or down, of course. Gliding from one deck to another, space scientists chose up or down as they wished or whatever seemed to suit the situation best. Indeed, it sometimes happened that crew members conducted a conversation with one's head up and the other's head down. Regardless of position, once in the new deck astronauts could look through the grill into the quarters they had just left.

The dome of the S-IVB led to the airlock, which could be sealed and used as a way station for space walks. Beyond the airlock was the docking module to which the Apollo ferrying craft was attached.

In the docking module was the massive control console for Skylab's solar observatory, itself a cluster of eight 10-foot-long telescopes known as the Apollo Telescope Mount (ATM). Five of the telescopes recorded x-ray and ultraviolet waves on film. A sixth created an artificial eclipse to record the Sun's corona. Two more photographed the Sun's surface.

Skylab didn't provide the luxury of *Star Trek's Enterprise*, but NASA still wanted to offer crews the good life, or as good a life as a space people could have in an orbiting fuel tank built by the lowest bidder. Of course, the good life according to NASA had its restrictions: song, but no wine or women. For song NASA provided cassette players, and each member of the crew brought his own tapes. But there were no wine and no women.

Pete Conrad was commander of the first Skylab crew. He had flown two Gemini missions and commanded Apollo 12, the second Moon landing. He advocated cocktails and, if not cocktails, at least beer.

Conrad tried a scientific argument on the white-shirted, dark-suited bureaucracy. Skylab was to be the space home away from home, set up to duplicate, as far as possible, the life of Houston, Clear Lake, Meadow Green, and the other towns where the astronauts lived. This was to be a complete medical experiment, designed to duplicate *every* aspect of normal American life with only weightlessness at medical issue. Then surely a conventional diet, including some semblance of a bar, was called for.

Try as Conrad might, he could not convince the top echelon of NASA into believing cocktails or even a little hops did not give the wrong image of the "right stuff." The Skylab crews were supplied with little more than a few bags of Tang orange drink, some instant coffee, and half a ton of nutritious Merritt Island swamp water for drinking and showers.

The shower was ingeniously designed as a round waterproof cloth enclosure fastened between ceiling and floor. Climbing inside, the astronaut attached a pressurized bottle filled with hot water and liquid soap to the ceiling above himself. These two liquid essentials were piped to a hand-held shower head. When the astronaut finished showering, the enclosure collapsed to the floor to be stored away compactly.

Safety was a weightier consideration than the amenities of the good life. Skylab was carefully insulated against solar radiation. The astronauts' golden work clothes were flame-resistant. Throughout the craft were 22 fire, smoke, and pressure-decrease alarms. Also aboard and studded throughout the workshop were sensors for toxic gases, meteor impact, and solar flares.

Alarms were placed next to the intercom system that had been installed for vocal communication, but feedback in the system created a continual whistle. Astronauts had to depend on shouting, and one of the health hazards of Skylab proved to be hoarseness.

On May 13, 1973, Saturn V blasted the unmanned Skylab from pad 39A of the Kennedy Space Center. Later, crews would rocket to Skylab, dock, then return to Earth in an Apollo Command Module.

Just 63 seconds into Skylab's unmanned launch a mission controller saw on his monitor that a shield protecting the workroom and living quarters had unexpectedly deployed and Skylab's thermal protecting insulation had been stripped away. The temperature of the space station began rising rapidly. It soon reached the level where plastic inside was melting.

Skylab's first crew was supposed to be launched the next day. Teamed with Pete Conrad, the commander, was Joseph Kermin, a medical doctor and Navy pilot, and, Paul Weitz, a Navy test pilot. Their launch had to be delayed.

Eleven days later Conrad, Weitz, and Kerwin were propelled into orbit in an Apollo module with 400 pounds of specially designed repair equipment and three masks to protect them against possible air pollution.

"We can fix anything," Conrad promised, and he fulfilled his boast. Working outside Skylab in space suits, the crew positioned a parasol over the living quarters and workshop. Almost as soon as they roped it securely, the temperature inside began dropping. Over the next week, it fell from 110 into the 90s, the 80s, and finally to the livable 70s.

The space station looked like a goose flying on only one wing, but hard work, ingenuity, and teamwork had paid off. The food supplies, though, were curtailed. Frozen steaks and desserts had stayed frozen. Dried foods, including dehydrated Lobster Newburg, was ready for festive reconstitution. Long exposure to intense heat, however, had spoiled all canned food. The next 26 days, even without their canned foods, Conrad and crew manned the station, and shot over 28,700 pictures of the Sun on the solar telescopes, shot about 10,000 pictures of the Earth, and recorded 8½ miles of scientific data on tapes.

Skylab's second mission, scheduled for a July launch, was to be a record-breaking 56-day medical marathon. Other goals included extensive solar observation and tests of a small, zero-gravity, materials processing factory.

Alan Bean, who had walked on the Moon on Apollo 12 with Conrad, commanded the second crew. With him were Owen Garriott, a professor of electrical engineering from Stanford, and Jack Lousma, a Marine test pilot and aeronautical engineer.

The TV star of the second flight, however, was not one of the astronauts but Arabella, a spider. She was yellow, with a big black spot, rather fuzzy, and came over on TV as cross-eyed. Arabella was part of an experiment designed to measure the adapability of animals, fish, and insects to weightlessness. What made her a celebrity was how quickly she adapted to living in space. The first web she spun was ragged and irregular. Yet on her next try she wove a beautiful, regular, geometric pattern, to the admiration of her TV audience.

The commander of Skylab's third crew was Gerry Carr, a Marine lieutenant colonel with a masters of science in aeronautical engineering from Princeton. Carr helped develop the four-wheel-drive lunar rover used on the last three Apollo missions and had tested it for the Moon surface by churning around the sands of Death Valley.

Edward Gibson, a solar physicist, was the science pilot, and the third member, Bill Pogue, was a former member of the Air Force Thunderbirds. Pogue was also a professor of mathematics at the Air Force Academy in Colorado Springs and before that a test pilot for the British Ministry of Aviation.

In the fall of 1973 Skylab's third crew was briefed for a special mission. In addition to their planned goal of making Earth observations and continuing the solar experiments, members of the third crew would also track and observe the newly discovered Comet Kohoutek from their unique perch in orbit.

Eight months before, during the nights of March 7 and 8, 1973, at the Hamburg Observatory, West Germany, Dr. Lubos Kohoutek was photographing an asteroid when he discovered he had photographed an object moving differently from the asteroids. Kohoutek recognized the object as a comet and, inspecting plates he took earlier, found he had first photographed it January 28 but had not noticed it before.

As do all discoverers of comets, Kohoutek wired the Central Bureau for Astronomical Telegrams at the Harvard-Smithsonian Center for Astrophysics. The Director of the Bureau, Brian Marsden, confirmed Kohoutek's discovery, and numbered it 1973 XXI, the twelfth comet sighted in 1973. Marsden named this visitor among the planets of the Sun Comet Kohoutek. This was no big deal for Kohoutek; he had given his name to comets before.

Aboard Skylab, the third crew would provide another giant leap for mankind. For the first time, scientists in space and above the distorting blanket of the atmosphere would directly observe and study a comet. That fact itself was an important contribution to science. In addition, 11 years behind Comet Kohoutek was the Earth's celestial clock, Comet Halley, sailing toward the Sun for its 30th recorded appearance. Astronomers all over the world hoped that, like Comet Kohoutek, Comet Halley would be studied by astronomers in space. According to Kohoutek's orbit, there was a little less than a month to prepare for the comet. All three crew members were rookies in space. The crews for Skylab had more careful and a longer preparation than any astronauts at that time, but nothing, not even the buoyancy tanks at Marshall and Johnson Space Centers prepared these rookies for living and working weightless.

Carr, Gibson, and Pogue, all scientists, understood thoroughly Isaac Newton's three laws of motion. Unfortunately, on Earth the laws don't operate with experimental purity. Forces like gravity and friction interfere. Not so in space.

Newton's first law says a body continues in a state of rest or in uniform motion in a straight line unless an unbalanced force acts on it.

When members of the third crew arrived in the space station, they were welcomed by three empty space suits in a state of Newtonian rest. Each suit retained the form of a human body, as if an invisible person were inside. The second crew had placed two of the suits seated at the table in the wardroom. A third was on the toilet. The suits, remaining in a state of rest, acted as a welcoming committee for the new crew.

As for moving in a straight line, once in motion, an astronaut couldn't change direction. It was 90 feet from the base of the S-IVB to the top of the docking adaptor. Only if astronauts aimed themselves perfectly was it possible to fly the 90 feet in seconds without crashing into anything along the way.

The flight path narrowed at the solar console and the airlock module beyond. All areas had boxes and spurs of various sorts sticking out. A bad aim might activate a sensitive switch by accident. These were painted red, but neutral protuberances were painted blue for a safe, if still painful, bash or for a grab and shove to provide the unbalancing force.

Gibson discovered that once he had floated away from a wall, he couldn't speed up his flight. Shoving against Skylab's wall joggled the spacecraft and affected the aim of the solar telescopes. So Gibson learned to shove off gently, so gently it could take 15 minutes to arrive where he was going. Or he might not arrive there at all but instead end up floating helplessly in midair until another astronaut powered him softly to his destination.

Gibson's gentle shove against Skylab's wall demonstrated Newton's second law: the acceleration of a body is directly proportional to the force exerted but inversely proportional to the mass of the body. Attempting to improve on Gibson's flight, Pogue transformed himself into a bird with paddles on his hands and feet and attempted to flap and kick his way across the space station. The flurry of flapping and kicking, however, influenced neither speed nor direction. Mr. Pogue's experiment would not amend Mr. Newton's laws.

Newton's third law says that whenever one body exerts a force upon a second, the second exerts an equal and opposite force upon the first. A sneeze threw an astronaut into continuing backward somersaults until he grabbed something. Using a wrench caused trouble unless the users' feet were anchored.

Without gravity to establish an up and down, all the crews,

and even the non-human passengers, adapted to an up and down direction by sight. The table in the wardroom, for example, was fixed to the floor and had an "up." The ergometer (the stationary bicycle) was also fixed to the floor, though Conrad sometimes operated it with his hands by floating above it. The solar console was oriented by a horizontal table at the base of a switchboard "above" it. Gibson could still operate the console from any direction but when he floated from "above," he had to adapt to "off is up" and "on is down."

Goldfish and minnows carried on board for an experiment decided that one side of their bowl, not the bottom, as it had been on Earth, was to be their down, and swam merrily in horizontal circles. Even newly hatched minnows accepted this majority decision.

It really didn't take long before everyone and everything aboard had adapted to zero gravity. All astronauts agree the condition makes for easier living.

It was November 26, two days before Thanksgiving in 1973. Comet Kohoutek had barely passed through the asteroids and crossed the orbit of Mars when Pogue spotted it through the window of the wardroom sparkling yellow, red, and orange. Gibson called it "one of the most beautiful sights in creation I have ever witnessed."

Skylab's scientific observations of Kohoutek were to be carried out using the telescopes of the Apollo Telescope Mount, whose primary mission was actually the study of the Sun.

The astronomers liked to work at the ATM's solar console. The console was isolated, the only place in Skylab that gave a sense of being alone. It was a small area compared with the living quarters and workroom. In addition, cassette players hung near the console, and each astronaut enjoyed his own choice of music during long observing shifts.

Carr turned the spacecraft so the ATM and the instruments inside could be trained on Kohoutek. Eight high-tech telescopes peered through the clear near-vacuum of space at the comet, giving these observers a significant advantage over even the largest telescopes on Earth, which have to peer through 80 plus miles of an atmospheric curtain.

One ATM telescope was a coronagraph supplied to NASA by the High Altitude Observatory on 14,000-foot Mt. Evans in Colorado. Coronagraphs use a small disk to create an artifical solar eclipse. Gibson, with the disk, blocked out the Sun and was able to photograph Kohoutek closer to the Sun than a comet had ever been photographed before.

November 28 was Thanksgiving, but aboard Skylab it was a workday. For the first six and a half hours, Gibson and Pogue went outside on a spacewalk (which the astronauts call EVA) to replace the film in ATM's solar telescopes. After the long EVA, all three were glad to be able to gather around the table for Thanksgiving dinner. Like all meals, this one was served on a tray, about four inches deep, much deeper than the usual cafeteria tray, so dishes were sunk in recesses. Each astronaut was his own cook, defrosting, reconstituting the dehydrated food with warm water, and mixing powdered drinks with cold water. The tray was fixed to the table and heated. Frozen and dehydrated food may not sound like that of the good life, but it certainly was an improvement over the squishy plastic envelopes used on the Apollo Moon trips.

Carr had a prime rib dinner, Pogue chicken and gravy, Gibson turkey and gravy. They had real forks and spoons, though utensils were magnetized to keep them on the trays. Liquids, of course, had to be in squeeze bottles to prevent clouds of coffee, milk, and cola from spreading and drifting throughout the space station.

Kohoutek put on its best show for the astronauts around Christmas. In addition to the Apollo Telescope Mount the astronauts had brought with them an electronic camera, a backup of one used on the Apollo 16 Moon mission. It was very adaptable, and they could use it either through the windows of the Apollo Command Module or outside.

The comet was now very close to the Sun and invisible in its glare, even when the astonauts looked through the darkened bubble of the space-suit helmet. Using the coronagraph, however, Gibson located the comet and gave its exact position to Carr and Pogue, who went outside to take pictures of their invisible subject. They pointed their cameras on the coordinates, clicked away, and made pictures of the comet when it was just a few degrees from the Sun, shooting out a beautiful tail and jets. Eventually the comet was so close to the Sun that these techniques didn't work any more.

On Christmas Eve, the families of the astronauts gathered at Johnson Space Center mission control. Aboard Skylab the astronauts used the portable TV camera to show their families the tree they had built out of food containers. The second crew had hidden Christmas presents around the spacecraft, and the flight director sent up hints on the teleprinter to help Carr, Pogue, and Gibson on their treasure hunt.

With the Skylab space station already in orbit, the Goddard Space Center for NASA, the National Science Foundation, and the International Astronomical Union combined their facilities to gather as much data as possible on the comet. If these institutions could work together on short notice and achieve results, astronomers believed that with 10 years planning they could structure an exciting and scientifically valuable Halley mission, one that included a space station more sophisticated than Skylab.

To verify Skylab's observations of Kohoutek, the Naval Research Laboratory at White Sands, New Mexico, launched an Aerobee rocket to check for a gigantic hydrogen cloud emanating from the water vapor driven off Kohoutek. The Research Council of Canada sponsored a flight of a Convair 990 aircraft to take high-altitude spectrographs, and Mariner 10, on its way to Venus, scanned the comet's ultraviolet spectrum.

Telescopes peered from the ground through the Earth's atmosphere as well. McDonald Observatory on Mt. Locke is 6700 feet high in the Davis Mountains of southwest Texas. No light pollution interfered here because the nearest town, Kent, is 30 miles away and consists of little more than a gas station and restaurant. McDonald boasts the darkest nights in the United States, and at night the brightest lights are stars or the Moon on its monthly traverse.

McDonald's director, Harlan Smith, and astronomer Ed Barker used four telescopes, a 30-incher, a 36-incher, an 82-incher, and their 107-inch giant, to study Kohoutek.

Kitt Peak National Observatory in the Papago Indian Reservation, Arizona, has the greatest concentration of telescopes in the world, and three of the largest were aimed at Kohoutek: an 84-incher, a 90-incher, and a 158-inch goliath. A battery of telescopes on Mt. Hamilton's Lick Observatory near San Jose, California, run up to 120 inches. Most of these were focused on the comet as well.

On Sugar Loaf Mountain in the Rocky Mountains near Denver, Colorado, a shivering Gary Emerson stood in snow up to his knees pointing a 9-inch telescope he had made himself at Kohoutek. During extreme winter cold, Emerson says an astronomer figures how many layers of clothes to put on by subtracting 10 degrees from the temperature. Emerson couldn't put on enough layers to keep warm. His subtraction put him at 20 below that December, in 1973.

Emerson's interest in astronomy started when he was 13 years old. He was living in Chicago and bought a new 3.5-inch skyscope for $30. He was a member of the Chicago Astronomical Association, which was tracking a Russian satellite from the roof of the Edgewater Beach Hotel. He eventually got to the Rockies as an astronomer at the High Altitude Observatory on Mt. Evans.

Now he set the motor drive of his 9-inch telescope so it followed Comet Kohoutek automatically. He thought wistfully of McDonald where the astronomers aimed their telescopes by computer from warm control rooms and watched the sky in comfort on television screens and computer monitors. Emerson and his wife, Paula, lived in a mountain shack warmed by a wood stove.

Emerson created his 9-incher from an Air Force camera a

supply sergeant tossed into a dumpster. It was one of four cameras originally made for wide-area aerial reconnaissance before satellites took their place. It weighs 80 pounds. Since its lens is a meter in focal length, Emerson had little trouble transforming the camera into a photographic telescope. He mounted it on the electric motor drive that can track a star or a comet moving across the sky (or actually while the Earth moves under them) for 2 hours without adjustment.

With his converted Air Force camera, Emerson mapped all the stars of the Northern Hemisphere, carrying on work E. E. Barnard began 90 years ago at the Yerkes Observatory, Green Bay, Wisconsin. One of the pictures Emerson took on this project showed 1.5 million stars. The pictures he took of Comet Kohoutek interrupted the mapping, but the comet was a worthwhile surprise.

While the crew of Skylab was photographing the comet, its discoverer, Lubos Kohoutek, arrived at the Johnson Space Center. Kohoutek was completely astonished that his comet was the subject of T-shirts, cloth hats, and quickie paperbacks illustrated with photos of 1910's Comet Halley. He was dumbfounded that every time he spoke a battalion of reporters, photographers, and TV crew members reported what he said.

The news department at Johnson hooked up Mission Control to the TV networks to overhear a conversation between Skylab and Kohoutek. Carr, Pogue, and Gibson spoke with Kohoutek for about 11 minutes, a minispectacular that ABC, CBS, and NBC cut down to 3.5 minutes for the six o' clock news.

During the last week in December 1973, Kohoutek was hired to lead a cruise on the *Queen Elizabeth II*. He was accompanied by Buzz Aldrin and Neil Armstrong, famous for their Moon walk. Hugh Downs, anchor on the *Today* show, was there with Carl Sagan, then a professor of astronomy at Cornell University.

Kohoutek, more at home in an observatory than on shipboard, was seasick the whole voyage. The voyage itself was a fiasco since the captain of the *QE II* could find no open sky.

Not long after the beginning of the new year, Comet Kohoutek vanished from earthly view for all but Skylab and the largest telescopes.

Astronomers figured this was the comet's first visit around the Sun and its surrounding planets. The comet was huge when Kohoutek picked it up near the orbit of Jupiter on his telescope, but it lost mass quickly as it neared the heat and power of the Sun. Mathematicians figured its orbit only as a parabola, an extended U, so this visit was not only the comet's first but also its last to our solar system.

The average sky watcher on Earth who was expecting a dazzling celestial spectacular was disappointed. A newspaper columnist renamed it "Comet Edsel." Fred Whipple, former director of the Harvard-Smithsonian Center and probably today's most famous cometary astronomer, warned everyone, "If you want a safe bet, bet on a horse. Never bet on a comet."

Back on Skylab, Gibson took his last photo of Comet Kohoutek February 1, eight days before crew members had to leave their home in space. They had already been in space 76 days. They had a record and wanted to extend it, but a couple of months before, just 16 days into the mission, one of the three gyroscopes that stabilized the spacecraft had stopped functioning. When the second gyroscope on Skylab froze tightly, Mission Control decided it was no longer safe to operate the craft. It was time to go home.

During the night of February 7, the teleprinter on board Skylab began clicking continuously. On the morning of the eighth, Carr found 15 yards of instructions spewed across the cabin, a checklist of things the crew had to do broken down by hour and minute:

19:29 Verify that differential pressure of AM (Airlock Module) is zero.
Open forward AM hatch.
Turn on AM lock compartment lights.

19:30 Engage hatch handle retainer pin.

Verify differential pressure across
aft AM hatch is zero.
19:31 Open Aft AM hatch.

And on and on across the workroom.

Throughout the 83 days Carr and the crew received their duty lists in this detail. They tried to limit the paper invasion of Skylab but never could. One day Carr and the others refused to perform the duties, a day recorded in NASA history as "The Skylab Mutiny." On this last day Carr was sarcastic: "I understand you're to teleprint the Old Testament tonight," he remarked to the capsule communicator (CAPCOM), who had sent the messages from mission control at the Johnson Space Center.

However, crew members dutifully followed the long checklist for their home voyage. When they completed the list, they crammed the Apollo command module with all their scientific data, and when some of the stuffed boxes wouldn't close, "Force them," said CAPCOM unscientifically.

The crew finally forced the boxes and everything else including themselves into the command module. Carr separated the module from Skylab after 84 days and 1213 revolutions around the Earth. As the Apollo Command Module moved away from Skylab, Crippen, the CAPCOM asked the crew to say goodbye. He added, "She's been a good bird."

On February 8, 1974, the astronauts splashed down off the Mexican coast and floated for a half-hour before the *USS New Orleans* picked them up. The spash-down of the module was the first such Earth return not televised to the public. That Saturday the networks thought they could not preempt such important shows as *I Love Lucy, Roller Derby*, or a live basketball game between Providence University and Seton Hall.

The three Skylab crews had brought back 230,000 pictures, 45 miles of magnetic tape, dozens of biological samples. Among them the three crews had photographed the world from Newfoundland to the southern tip of Chile, from Mongolia to New Zealand. From these photos new deposits, especially of copper and oil, were discovered. And the astronauts discovered pollution at sea in both the Atlantic and Pacific.

From the Skylab stay doctors discovered that humans could adapt to weightlessness for periods of months. Pioneers have always adapted to the alien environment of their respective frontiers. Pioneers in the alien environment of space are no different from their forebears, and like them the astronauts all spoke of a "spiritual" change as a result of seeing their new frontier—a view of the world from the vantage point of 270 miles in space.

They saw one world, a world without political boundaries, and Gibson, looking outward, recognized the probability of other worlds in space.

"Seeing your own Sun as a star makes you realize that the universe is big and just the number of possible combinations which can create life enters your mind and makes it seem much more likely."

The Kohoutek data the astronauts had gathered were sent to leading cometary astronomers throughout the world, including John Brandt at Goddard Space Flight Center in Maryland, Fred Whipple at Harvard, and Don Yeomans and Marcia Neugebauer at the Jet Propulsion Laboratory in California. The scientists punched the data into their computers, compared data, and checked theories. They found some new answers about comets in the computer printouts. However, the new data still left questions to be answered.

Although Carr, Gibson, and Pogue had been able to photograph Comet Kohoutek across the near vacuum of space, their cameras had not been able to penetrate the coma, that atmosphere of gases and dust that surrounded the nucleus. They had not been able to see into the heart of the comet.

The NASA Office of Space Science decided to bring the cometary astronomers together at the Goddard Space Flight Center during the fall of 1977 after the scientists had time enough to evaluate the data gathered on Skylab. What did this new body of knowledge, the first ever obtained above the Earth's atmosphere, reveal about the composition of comets? What did the data reveal about the origin of comets, about the origin of the Earth and the entire solar system? How did it relate to what they already knew, or thought they knew, about comets?

Scientists and engineers would also be able to plan future investigations. McDonnell-Douglas had prepared a second S-IVB, a duplicate of Skylab called Skylab B. They promised NASA it would be ready to welcome Comet Halley at the end of 1985 or the beginning of 1986. A flight through space to Halley presented many problems. If a spacecraft, manned or robot, reached a comet, would it be possible to send it into the coma to analyze the nucleus close at hand? Would the astronauts or the spacecraft survive the fusillade of particles shot out of the nucleus by the power of the Sun? At the conference, the scientists and engineers would have some real mysteries to ponder.

3

High on Mercury Ions

During the second week in October 1977, the top cometary space scientists were on their way to the NASA conference, Space Missions to Comets. Their cars turned into the main gate of the Goddard Space Flight Center, Greenbelt, Maryland. The guard at the visitors' parking lot, wearing white gloves, a whistle on a golden chord around his neck, seemed to have learned his technique watching Leonard Bernstein on TV. He could be guiding the Philharmonic through Prokofiev's "Love for Three Oranges" rather than directing cars to their places between the yellow lines.

It had been four years since Gibson, Carr, and Pogue had brought the tapes and pictures of Comet Kohoutek down from Skylab. Since then some of the astronomers arriving here had coordinated material collected on Skylab with material accumulated by McDonald Observatory, Kitt Peak, and many other sources. Their computers compared, checked, extrapolated, and evaluated the spectroscopic, visual, and radio data, but there was so much information that by the time of the conference, they had reduced only about half of it to a form they could study.

Scientists and nonscientists (including Washington administrators and politicians) gathered in the auditorium at Goddard. Fred Whipple's proposition that "dirty snowball" best described the composition of a comet had been greatly strengthened by the Skylab observations. For the conferees in the auditorium at Goddard he projected a slide of Kohoutek that Carr had taken back on Skylab. The bright globe of the coma, about 360,000 miles across, hid within its blaze the unseen nucleus.

Behind the coma two tails radiated for 10 million miles across the sky. One tail, the dust tail, was slightly curved, and golden in color.

Whipple explained that the dust was thrown off the nucleus by the pressure of escaping gases warmed by the Sun. Comet dust is not the stuff you find under the bed of a badly cleaned house but, on the contrary, consists of tiny particles a few tenths of a micron (a millionth of a meter) in size shot out at more than rifle velocity.

The other tail was the ion tail. For the nonscientists, Whipple described ions in some detail because they are important in understanding comets. The nuclei of atoms have an equal number of electrons and protons and are therefore neutral magnetically. When the solar wind, itself magnetic, drives these atoms from the head of the comet into the tail, it knocks off electrons. When it loses an electron, the atom is positively charged and is now called an ion.

The ion tails of comets appear blue in photographs because of the strong emission of blue light from ions and molecules like cyanogen. Indeed, this fact was known even when Comet Halley appeared in 1910, and some persons panicked and feared the poisonous cyanogen would kill earthlings as the comet's tail brushed the Earth May 19 that year. The gas killed no one. Percival Lowell, the astronomer noted for believing there were canals on Mars, described the tail as "the airiest approach to nothing in the midst of naught."

The word *plasma* is sometimes used to describe the ionized gas on the ion tail; the word sounds more esoteric than it is. Plasma is found in fluorescent tubes, for example. The plasma of neon glows reddish-orange, seen often in neon beer signs.

In the coma and tail of Kohoutek, spectrometers had identified ionized atoms of carbon, hydrogen, oxygen, and nitrogen as well as molecules of water, carbon monoxide, carbon dioxide, and hydroxide. These and other ions are called the "daughters" of unseen "parents" in the nucleus. Astrophysicists have identified the ions of daughters in the coma and tails by those Fraunhofer lines on the spectrum. However, they have not been able to identify the parents in the nucleus of the comet, hidden inside the coma.

If the nucleus of the comet was ice and dusty dirt, as Whipple predicted 30 years before, then the Kohoutek data should indicate a great cloud of hydrogen gas, which would have been created by the Sun breaking up the water molecules, extending from the coma.

On the screen Whipple showed a slide taken in ultraviolet light by one of Skylab's telescopes in which the hydrogen cloud was clearly visible. On the cloud, an artist superimposed a circle the comparative size of the Sun. The comet's hydrogen cloud extended beyond the circle of the Sun as if a nickel lay on a large ashtray. Whipple called this picture one of his favorites. The photograph not only confirmed the presence of water-ice in the nucleus, but demonstrated just how impressively large comets can get as they are heated in their approaches to the Sun.

But Whipple warned the conferees:

> Comets are the greatest little deceivers in the solar system. A tiny body puts on a magnificent show by ejecting vapor and particles so that the solar radiation reflecting on the particles and being re-

radiated from the gases produces a conspicuous comet. A five to ten-kilometer body can produce phenomena that stretch out visibly over a hundred million kilometers or more.

The Skylab instruments and the Kitt Peak radio telescopes found other gases in this cosmic deceiver: methyl cyanide and hydrogen cyanide, gases carrying nitrogen, oxygen, hydrogen, and carbon, the fundamental gases of the Earth's atmosphere.

This evidence further confirmed that comets really could be Rosetta Stones through which we could learn the secrets of our cosmic past. Further investigations of other comets might be able to tell us how this, the third planet of the solar system, has its special life-giving water and atmosphere.

We know the Earth once had a primeval atmosphere probably consisting mostly of hydrogen but that the gravitational force of the Earth in that era was too weak to hold on to the atmosphere. Did comets colliding with the Earth replace that lost atmosphere? Can our current atmosphere be what remains from early comet storms occurring between 2½ and 3 billion years ago?

The Earth has a circumference of 24,800 miles, gigantic compared with a comet's nucleus, which is about only 6 miles wide by 10 long. Still, the nucleus of a comet is larger in bulk than Mount Everest, and 10 Mount Everests crashing onto the surface of the Earth would certainly transform the planet.

Imagine a "comet storm" with thousands of Everests crashing down on Earth. The frozen water-ice of the comets' nuclei would thaw from the impact. Water would fill the basins of the lava crust in the Earth's surface forming blue seas. The gases released would replace the lost primeval atmosphere.

This picture of the atmosphere and seas brought by comets to the primeval Earth is only a speculation, but it is one that has been growing in acceptance. So far, the tangibles in the scenario are the seas, the atmosphere, the comets, 12 or so that visit us each year.

Within the last 200 million years, relatively modern times,

geologically speaking, celestial objects crashing on the Earth's surface have left about 70 impact craters like those near Winslow, Arizona, and Manicouagan, Quebec. These craters were formed when the orbit of an object in space crossed that of the Earth, and as a result of the collision the object plowed deep into the ground. That object might have been a comet, a fragment of a comet, or an asteroid jerked out of the asteroid belt between Mars and Jupiter by some gravitational tug.

It is now being theorized that the crash of a comet, a huge meteorite, or perhaps an asteroid killed off the dinosaurs 65 million years ago. The collision flung a blanket of dust into the atmosphere that blocked the sunlight and lowered the Earth's temperatures. This "cometary winter" not only killed the dinosaurs, and many other species as well, but eradicated their food supply.

The theory that a collision between the Earth and an object from space killed off the dinosaurs was not proposed until a year after this conference at the Goddard Space Flight Center. Scientists have recently discovered that this collision, which occurred about 150 million years ago, left behind two substances rare on Earth but plentiful in whatever it was that struck our planet: first, a thin layer of iridium, a chemical element rare on Earth but plentiful in meteorites; second, within the layer of iridium, amino acids.

Certain amino acids are necessary for the first evolutionary step toward life. During the visit of Kohoutek, scientists both in Skylab and in laboratories on Earth hoped their spectrometers would find ammonia and methane in the gases of Kohoutek's coma. The presence of these two gases would indicate the existence of amino acids within Kohoutek's nucleus. If this were so, then the case might be made that comets could have brought to Earth not only the material for our seas and atmosphere, but also the first elements of life.

Alas, neither methane nor ammonia were found in the coma of Kohoutek. However, Armand Delsemme, professor of physics and astronomy at the University of Toledo, who addressed the conference, said that if it was found that parent molecules in the nucleus of Comet Halley showed evidence of elements rare on Earth, such a discovery would be a scientific event as revolutionary as Darwin's theory of evolution, proposed in 1859.

Astronomers studying the comets have a difficult time compared to, say, archaeologists. For example, archaeologist Donald Johanson was able to bring the bones of little Lucy, the 3-million-year-old Australopithecus, from Ethiopia to Cleveland to study them. Astronomers can use only the light sent to them through space from some distant object, possibly millions of miles away.

A valuable space mission, Delsemme suggested to the conferees, would be to have a spacecraft penetrate Halley's coma and close in on the nucleus. Only in this way could the Rosetta stone be sampled.

Kenneth Atkins, Senior Engineer at the Jet Propulsion Lab, Pasadena, California, had good news for Delsemme and the other astronomers. He told them JPL was developing a way of getting to the comets. Referring to Delsemme as a Captain Ahab going after the "great white ghost in the sky like Moby Dick," Atkins promised Delsemme a harpoon. The harpoon was actually a mercury ion engine, already developed by Harold Kaufman at NASA's Lewis Research Center, Cleveland, Ohio. The propulsion of the engine is fueled by mercury ions. Atkins projected a slide of this new Solar Electric Propulsion System (SEPS) on the screen. From the picture the engine looked something like a bass drum with one end covered with a double, wire-mesh screen.

SEPS' simple appearance was deceptive. A very high-tech, state-of-the-art engine, SEPS promised to be a great advance in the field of space exploration. It uses its fuel so efficiently that it can outdistance even the best conventional propulsion system of Wernher Von Braun and his colleagues, who developed the first U.S. rockets.

Inside the engine, the mercury is vaporized by intense heat, and the gas that results is pumped into a chamber where the vapor is ionized, then accelerated. The engine's ion accelerator is driven by a solar power system with wings 12 feet wide and 450 feet long. The solar cells are very efficient, each equipped with a concave reflector designed to focus and intensify the sunlight. The accelerated mercury ions then blast out of a nozzle. While the engine starts slowly, by accelerating at a constant rate the spacecraft can be flying about 1140 miles per hour in 30 days, and in three years it can be shooting through space at 43,200 miles per hour. It then can continue on its way through space for a very long time. The system is flexible. Each bass drum is a unit on its own, and several can be clustered. Atkins said that JPL was planning to use four with two more as a backup to propel a spacecraft of about 4 tons. In 1977 the space shuttles had not yet flown, but contracts to develop the orbiter, the rockets, and the boosters had been assigned. As America's primary launch vehicle, the shuttle would replace the large expendable rockets, the Atlas and the Saturn, and deliver all NASA payloads into low Earth orbit. Atkins foresaw that SEPS would provide NASA with its second step. In clusters of four with two as backups, SEPS could power spacecraft from the shuttle to comets, planets, and asteroids. One project, the Venus Orbiting Imaging Radar (VOIR, now renamed Magellan) was designed to circle Venus, penetrate its cloud cover by radar, and map the entire planet. JPL believed both the Halley mission and VOIR could be powered by SEPS.

SEPS was the good news the cometary astronomers had come to Goddard to hear—a practical means of accomplishing a Halley mission, a mission that would accomplish more than flying through the coma. Comets move so fast that no chemical rocket can match their speed. But SEPS could, and it made a rendezvous with Halley a real possibility.

The leaders of the comet community, Brian Marsden, Director of the Central Bureau for Astronomical Telegrams at Harvard, Whipple, Delsemme, Neugebauer, and others, all wanted to send a strong message about the feasibility of a scientifically advanced Halley mission to the highest echelons of NASA, especially to William Lilly, NASA's Associate Administrator for the budget.

When the general presentations in the auditorium were finished, the conferees broke up into workshops and met with Thomas Young, the Director of NASA's Planetary Studies Division and with two others from NASA's headquarters. The scientists wanted to know if NASA had provided for a comet mission in their requests for 1978–1979 to President Jimmy Carter's Office of Management and the Budget (OMB).

Young did not reveal the details of the budget proposals, because, he explained, the budget was secret until the president presents it formally to Congress. However, Young was happy to report that the proposal that would go to Congress in January 1978 included funding for the research and planning for a Halley mission. The congressional budget decision lay with four committees, one in the House of Representatives, one in the Senate, and two special subcommittees, that all had jurisdiction over space. Although strong interest in Comet Halley was still four or five years away, Young promised that NASA would impress on the OMB and members of Congress the necessity of getting an early start. To get SEPS engines built and tested for a 1984 launch and a 1986 rendezvous with Comet Halley required an immediate go-ahead.

By the time the conference had broken up, the good news had overtaken everyone. The conductor of the parking lot moved the cars out with his usual virtuosity, but to the conferees he seemed to be conducting Prokofiev no longer. He was now conducting Beethoven's "Ode to Joy."

Skylab was still floating 200 miles above the conferees who were dispersing toward their various home bases. However, Skylab was a silent ghost, its power shut down. It was tracked only intermittently and almost forgotten. NASA planned to boost it into a higher orbit and keep it flying until a space shuttle could rendezvous with it to repair it and resupply it. The budget actually included funds for this mission.

The North American Air Defense Command (NORAD) went into action and aimed a radio signal at the orphaned spacecraft. After five years Skylab heard the voice of Earth again. Skylab reported on its own condition, but transmission, powered only by solar cells, stopped as soon as the craft went into the Earth's shadow. NORAD got two batteries charging and powered to maintain continuous signals.

At the Johnson Space Center controllers again authorized the mission control monitors to attempt restarting the space station. One of its three gyroscopes began to spin slowly and jerkily. In spite of this start, Johnson mission control could not turn the old ship into a livable home in space.

Suddenly a Russian satellite, Cosmos 954, flamed into the atmosphere of the Canadian Arctic and came down scattering radioactive chunks all over the snowy wilderness. Newspapers and TV covered the debacle in detail and even reminded an alarmed public that Skylab was still up there, 100 tons of it that might also fall from orbit and hit the atmosphere, turning into a free-falling missile.

Cartoonists made drawings of Skylab, like Cosmos 954, flaming down on Omaha, Nebraska, or Corpus Christi, Texas. The departments of State and Justice demanded hourly reports from Johnson Space Center. Foreign embassies sent third secretaries scurrying to NASA headquarters asking that Skylab be directed away from their skies.

Although the United States assured everyone that Skylab was secure in its orbit, it turned out that the orbit was suddenly deteriorating faster than anyone had anticipated. At the same time, the shuttle launch schedule was slipping badly. No orbiter could get up there in time to rescue Skylab. NASA decided it had to wish Skylab a mournful but careful goodbye by having it burn up in the Earth's atmosphere under their control.

The controllers moved Skylab away from populous areas and toward either the Pacific or Atlantic Ocean. Were Skylab's 100 tons to hit the atmosphere in one piece, the Earth would have the spectacle of a roaring, blazing meteorite whose shock wave would level a small city and sprinkle the debris with blazing chunks of metal. An apprehensive world awaited the outcome of these maneuvers.

In Washington, DC, a group of imaginative computer operators organized Chicken Little Associates: send them your general location and, for a price, they figured out your odds of being hit. The Psycho-Energetic Institute contacted all persons known to have ESP and asked them to concentrate together on redirecting the orbit of Skylab into the Pacific. The radar at NORAD detected no discernible change in the path of the craft. The San Francisco *Examiner* offered $10,000 to the first person who brought an authentic piece of Skylab to their office.

A few minutes before 1 P.M. on July 11, 1979, NORAD received reports that people had seen flaming objects falling into the Pacific Ocean off the south coast of Western Australia. More objects were seen flashing into the outback of the Great Victoria Desert. Others heard sonic booms and the high-pitched whirling sounds of unseen somethings shooting through the air.

The personnel at NORAD and at Johnson Space Center waited anxiously for reports of damage. None came in. No kangaroo, no jackrabbit, no dingo had been touched. Stanley Thornton, a beer truck driver from Esperance, a small port town on the Western Australia coast, found a charred piece of something in his backyard. He flew with it to San Francisco without a passport. The *Examiner* had to rescue him from Customs and Immigration, but the reporters identified the charred something as a remnant of Skylab insulation. They gave him his $10,000. The Australian government filed a strong protest with the US government against being showered by dangerous NASA debris. The US State Department apologized.

Now public pressure prevented the use of Skylab B as a space station for the Halley mission. The pressure came from both a fearful public and the highest echelons of NASA. Robert Frosch, the NASA administrator, knew that Skylab B was too large to be carried in a shuttle cargo bay. Still intact, Skylab B was trucked to Washington, DC, and set up as an exhibit in the Smithsonian Air and Space Museum, where it can still be admired.

Two SEPS prototypes were manufactured by Hughes Aircraft and tested at the Lewis Research Center where they performed flawlessly. In its budget request for 1980 NASA asked for an appropriation to construct many more. Spacecraft on the Halley and other missions could be transported 200 miles above the Earth on the shuttle, then propelled by SEPS deeper into space.

4

No to SEPS, to HIM, and to HER

The vision that the Solar Electrical Propulsion System could bring to reality was harbored jointly by scientists of NASA and of the European Space Agency (ESA): not only a mission to Comet Halley but also a trip to a second comet that would take a spacecraft three astronomical units, about 280 million miles into space, a trip of more than 1000 Earth days.

This was the projected Halley-Tempel 2 mission.

NASA, the Comet Science Working Group, as well as ESA believed that a visit to Comet Halley amounted to only half a comet mission. The other half would be for the spacecraft to continue on into space and intercept a second comet. Halley is a relatively young, active comet with a dynamic atmosphere. The working group thought the data that experiments obtained from this comet should be contrasted with what could be learned from one older comet. An older comet would have spent more of its lifetime within the relatively warmer regions of the solar system, made more trips around the Sun, and thus lost more of its gases and dust. Joe Veverka of Cornell's Radio Physics and Space Research Laboratory, chairman of the Working Group, said:

> Without a second comet rendezvous, the Halley flyby alone is a good stunt but a scientific dud—mostly a mass of high quality data which would be uninterpreted until such time that a comet nucleus is studied adequately.

While engineers were testing the SEPS prototypes, astronomers at JPL compared the orbits of selected comets on their computers. Among many short-period comets, those whose orbit was less than 10 years (Encke, or Giacobini-Zinner, for example), they decided on Tempel 2, a comet with an orbit of a little more than five years.

Then the scientists designed the following mission scenario: two spacecraft, one encapsulated inside the other and both carrying state-of-the-art instrumentation, were to be deployed from a shuttle in July 1985. Then the outside craft's SEPS engines would take over and power the double craft to reach Halley by November 1985 when the Comet would be 75 million miles from Earth (about 10 weeks before it would reach perihelion).

Here the spacecraft would release its encapsulated probe, then continue on its way. The probe, ESA's contribution to the mission, would dive through the coma to within 1000 miles of the nucleus, its instruments signaling their findings to receiving stations on Earth as long as possible. The probe was called a flyby mission.

The mother craft would continue to sail on through space arriving at Comet Tempel 2 in 1988, three years later. The speed that SEPS would have built up to by then would allow the spacecraft to accompany the comet, flying beside it for a year, as close to the nucleus as possible, a sophisticated "rendezvous" mission.

The scientists who designed this mission hoped that their

probe might even land on the nucleus and stay there, thus becoming an explorer that would send back information on the composition of the surface and preparing the way for a lander in the twenty-first century that would be able to return samples of the Rosetta-Stone nucleus to Earth.

This was the vision, the dream made possible by the development of SEPS. It was an expensive dream, a $700 million project, though some of that expense would be shared with ESA.

ESA also shared NASA's enthusiasm for the Halley-Tempel mission. The Science Program Committee of ESA met at the European Space Operations Center in Darmstadt, south of Frankfurt, West Germany, to design the probe.

ESA is a consortium of 11 nations: Belgium, Denmark, France, West Germany, Ireland, Italy, Netherlands, Spain, Sweden, Switzerland, and the United Kingdom. Austria and Norway are associate members, and Canada has a special working arrangement with the agency.

Financially ESA is organized differently from NASA. ESA does not finance experiments out of its own funds. If Belgian scientists and engineers develop an experiment for a spacecraft, it is financed by Belgium. For the ESA directorate to suggest that something might have to be cut back or eliminated from a project would be an affront to Belgian pride. And there is French money and French pride, German money and German pride, Italian money and pride, and so on, 11 traps for their director of science, Roger Bonnet. Bonnet is a scientist with a solid reputation for spectroscopy. His job as director, however, requires more in the way of diplomatic than scientific skills.

ESA and its predecessors had already been cooperating with NASA on various programs for five years. Early on, ESA had put together a rocket of its own, with a British first stage, French second stage, and German third stage, appropriately called Europa. It was launched from Woomera, Australia. Then a French company had a rocket called Ariane in the testing stage, something less heterogeneous than Europa.

ESA sent out requests all over the world asking scientists to submit designs for experiments to be carried on their probe into the dust and gases of Comet Halley.

Late in 1979 Alan Delamere at the Aerospace Systems of Ball Aerospace Corporation in Boulder, Colorado, received a telephone call from the JPL. Delamere was an adviser to ESA for the Halley-Tempel 2 mission. He had earned his Ph.D. in engineering from Liverpool University in England and specialized in designing and developing precise scientific instruments. The telephone call Delamere received was from G. Edward Danielson, a researcher at the California Institute of Technology, who with Michael Malin at JPL, had designed a camera for the ESA probe. Did Delamere think there was any possibility of getting it on board? Delamere thought it was a great idea and called in Harold Reitsema, a young planetary scientist at Ball Aerospace with whom Delamere had worked on other projects.

Delamere brought the team's camera design to Bonnet's attention. Decisions at ESA are committee decisions, and a committee at Paris decided a camera would be a welcome addition if it could get close-ups and if the Americans could design a camera light enough and small enough that its inclusion did not necessitate removing one of the other instruments already accepted for the probe.

The four Americans made delicate modifications and came up with a new camera according to ESA's parameters. The committee in Paris liked the new design, and the camera joined the European instruments for the ride on the SEPS taxi to Comet Halley.

One evening in December 1979, Robert Frosch, the NASA administator, was, as usual, still at his desk on the seventh floor of NASA headquarters. He had just learned that the Office of Management and Budget had eliminated the appropriation for the construction of the SEPS engines from the NASA budget for 1980–1981 but retained a Gamma-Ray Observatory, also to be launched on the shuttle, but which still had construction woes.

Frosch exploded. While the Gamma-Ray Observatory was an enormously worthwhile project, he had not meant to set up an either-or situation in the NASA budget request. SEPS was such a tiny item in the NASA budget, and the total NASA budget itself had been averaging less than one-half of one percent of the total national budget. Eliminating SEPS had effectively eliminated the Halley-Tempel mission.

Frosch and Alan Lovelace, Deputy Administrator of NASA, visited the appropriations committees and the subcommittees and explained to the members of Congress exactly what SEPS was and what it meant to the US space program. They tried to get SEPS restored to the budget. They failed.

What remained in the budget and what was deleted had been determined by financial and other factors. Professor John Logsdon of George Washington University, the historian of the politics of the space program, explained that JPL could never have gotten three multimillion dollar missions in a row, one a year.

First, in 1978, the Galileo mission had been approved. This sent an unmanned spacecraft on a trip to Jupiter, through its atmosphere, and then on a 22-month, 10-orbit tour of the Jovian moons. In 1979, the Hubble Space Telescope was approved. Once deployed from a shuttle above the Earth's atmosphere, the telescope would be able to see to the edges of the solar system. Logsdon called the belief that the Office of Management and Budget would approve a third mission on the level of these "financial hubris" on the part of NASA.

In addition the 1980–1981 budget extended from the Carter administration into the Reagan administration. In Washington a change of administration is always a time of procrastination. Byron Johnson, former representative from Colorado on one of the House appropriation subcommittees, in his book, *The Bureaucratic Syndrome*, aptly described this constipated period as "a time when nothing passes."

At JPL it became black coffee and high speed computer time. Tim Mutch, NASA's deputy administrator for space science, told JPL scientists they had to drop SEPS and modify the

Halley-Tempel mission for Comet Halley alone and one that could be deployed from a space shuttle that the National Space Policy Board had declared would be NASA's primary launch vehicle to replace all the big rockets.

After coffee strong enough to launch a Saturn V, after thousands of silicon chips performed trillions of operations (and the computer printouts, if laid end to end, would stretch from Pasadena to Comet Halley), the planners at JPL created the Halley Intercept Mission (HIM). HIM was to be a spacecraft created as Skylab was out of spare parts: these from the Voyagers, Vikings, and Galileo, a Jupiter orbiter. After deployment from a shuttle, HIM could be powered to the comet by a liquid, two-stage rocket known as Centaur.

The strong coffee, supercomputers, and human imagination had created a new mission, though one nowhere near as potentially rewarding as the Halley-Tempel mission—no rendezvous but a flyby of Halley. For NASA, however, it was the only game in town and a lot cheaper. It was reviewed by members of the Comet Science Working Group, who declared their support for it.

During any period of budget cutting, funding is allocated by priorities. For science projects, including those of NASA, the recommendations are made by the scientific establishment, the US National Research Council, the National Academy of Sciences' Space Science Board, and others. The Comet Science Working Group is only one of these boards, and in the fund allocation feeding frenzy, project HIM came in second to the Venus Orbiter Imaging Radar mission (VOIR). The establishment thought that VOIR would be more valuable scientifically.

They remembered that earlier the comet group had called the Halley flyby "a good stunt but a scientific dud." The scientists didn't recall that when the comet group had said that, it was because the cometary astronomers were all pushing for the Halley-Tempel double mission. Now the Hally flyby, HIM, was all that was left for a Halley mission.

The comet group discovered that priorities couldn't be

changed either at NASA headquarters or in the President's Office of Science and Technical Policy. The group and JPL decided to whip up strong public support. They believed that the flight of Comet Halley over Earth would be the astronomical spectacular of the 1980s as it had been in 1909 and 1910. Already talk about Halley was on the TV evening news. Many public TV stations across United States were broadcasting a show hosted by Fred Whipple explaining comets. Nigel Calder beat the competition and had his book, *The Comet is Coming*, already in the stores. Bruce Murray hoped the leonine roar of public opinion would force a reversal of priorities. But it didn't. Washington remained impervious to the Halley din.

Cometary astronomers then decided to try to make HIM a private mission. They started a drive for donations at Cocoa Beach, home of Cape Canaveral. Not enough was collected. The cometary scientists had a friend in the President's policy office. She tried to influence the IRS to add a check box on tax returns. Check the box, and a dollar of your taxes goes to a Halley mission. But the IRS said, "No box."

HIM never got off the ground. However, HER almost did. HER was the acronym for Halley Earth Return, a mission proposed by Robert Farquhar, a senior engineer at Goddard and Director of the Flight Dynamics Division. The acronym is an example of Farquhar's sense of humor; however, it was not really meant as a joke but as a challenge to the space hierarchy.

In his office at Goddard, Farquhar is a relaxed and congenial man usually wearing a white shirt open at the collar with a pocket full of ballpoint pens, the very picture of an engineer. However his office is without that special kind of chaos that typifies the offices of most space scientists. Bookshelves run along one wall, well stocked but with room for more. The piles of offprints on his desk and table are unusually neat. His office lacks another piece of space-science equipment, the computer console and keyboard.

"Computers can't think," Farquhar explains. "I'd rather do my own thinking."

Farquhar likes solving problems. HER was a solution, a way to sample the gas and dust of Comet Halley as the spacecraft flies through its coma. The speed of the spacecraft would entrap the gas and dust through microscopic holes in Teflon tubes deployed out of the body of the vehicle. Inside the tubes the gas and dust would be held on an absorbent surface, a complex, sterile flypaper used to collect cosmic dust during flights of high altitude planes like the U-2.

The gas and dust, carefully preserved, could be released in a vacuum chamber on Earth for analysis. It was cheap and novel, and it might just revive the US chance at Halley.

When NASA received Farquhar's proposal, James Beggs, the Reagan NASA administrator, was enthusiastic: this mission was no scientific dud. This was the sample-return mission the scientists at the Goddard conference had dreamed about. This was bringing the bones of Lucy back to the lab. The bones of Lucy were only 3 million years old, but the gas and dust of Comet Halley were 4½ billion years old. The dust and gas samples could be as important as the rocks brought back from the Moon.

Beggs asked JPL to begin designing a spacecraft for the HER mission. By the time these designs were finished, NASA realized a spacecraft of this complexity could not be constructed in time to rendezvous with Comet Halley.

Usually Farquhar is most persuasive when someone says, "It can't be done." To NASA's surprise Farquhar quietly accepted their rebuff. They didn't know Farquhar already had another idea for a comet mission bubbling in his head.

HER almost saved a Halley mission for the United States, but almost isn't half-way. Andrew Stofan, acting associate administrator for space science in NASA, was ordered to notify Murray at JPL to stop all work on any Comet Halley mission and to refrain from starting any new ones. An official letter to this effect was sent from his office in September 1981. The United States, the world's leading spacefarer, would not be going to Comet Halley.

5

Halley: Ja, Hai, and Da

Almost as soon as the NASA letter notifying Murray to stop work on any Halley projects went to JPL, ESA knew it had lost its taxi to Comet Halley. For the first time the American National Aeronautics and Space Administration had reneged on its word on a major mission it had agreed on with its European counterpart. Representatives of the 11 nations of ESA gathered in Paris to decide how ESA should proceed.

Should ESA, like the United States, be content to watch from the ground as Comet Halley shimmered above? Or should ESA undertake a Halley mission on its own? If ESA did continue, it would have to develop a spacecraft different from the probe it had designed for the Halley-Tempel mission. And a new spacecraft meant newly designed instruments. As scientists pondered whether to continue, each nation also wondered if, for this first European deep-space mission, money from national budgets would be forthcoming. Comet Halley had glamor, of course, but how much rubs off on the nation financing the dust detector or a magnetometer? The ESA representatives, however, took a chance. They voted 12 to 1 for a European Halley Mission.

After the vote, The Scientific Program Committee of ESA now discussed just how it could put the mission together. The committee was made up of scientists from all over Europe—French (although they had voted "no," they acquiesced with the decision once it was made), West Germans, English, Belgians. Others, who were not even members of the the European Space Agency, came to help too. Among these were Roald Sagdeyev from the Soviet Space Research Institute, Fred Whipple from the Harvard Smithsonian Center, and Kunio Hirao, director of the Planetary Science Research Division of the Institute of Space and Astronautical Science at Tokyo.

The plan that evolved for the ESA mission was based on the following: in 1978 ESA had launched GEO-2, a spacecraft that studied the Earth's magnetotail. GEO-3, an improvement on GEO-2, was almost ready for launching. The committee adopted a plan in which GEO-3 would be modified for the Halley mission.

Bonnet also began negotiations with NASA to use a Delta rocket as a launch vehicle. NASA put the request on hold, so instead ESA committed its GEO-3 adaptation to its French rocket, Ariane. NASA's offer of a Delta, when it finally arrived, came too late to be accepted.

GEO-3, which had to be redesigned, was given a new, more poetic name, Giotto, after the thirteenth century painter, Giotto di Bondone. Giotto had painted a series of murals on the walls of the Scrovegni Chapel in Padua, Italy. In one of the murals he portrayed the three wise men presenting gifts to the Christ child. A comet dips down right over the stable, indicating this was the celestial object that guided the three wise men there.

In the centuries before Giotto's mural, comets were always portrayed as something like flowers that left behind them streamers depicted by various undefined geometric stripes. Giotto certainly remembered the visit Comet Halley made to the world in 1301 when he was 35 years old, and his painting is the first advance toward a modern, realistic representation of a comet. Giotto's comet in the mural has an easily recognizable nucleus and a fiery tail with streaks of flame and sparks. The tail leaves behind it a consumed substance, black against the blue sky. With ESA's help Giotto's name was going to be joined with Comet Halley again after 670 years.

The design of the new Giotto spacecraft inherited some of the characteristics of its predecessor, GEO. Giotto kept the shape of GEO, a squat barrel a little more than a yard in height and a tad over 2 yards in diameter. The power for the instruments would come from solar cells that formed the outer shell of the barrel.

Like Skylab and many satellites, Giotto would be controlled in space by jets and gyroscopes. Giotto's jets, using compressed hydrazine, were fired by ground controllers. Like GEO, Giotto spun at 15 revolutions per minute to maintain stability. Still revolving, it would dive right through the coma of Halley and within 300 miles of the nucleus, then out of the coma again. It would come closer to the soul of a comet than any man-made object had been before.

As Giotto was being adapted from GEO, it also took on many modifications for its new mission. The new design included shielding to protect Giotto and its instruments against the coma's super-rifling dust particles and the magnetic buildup of ions at the front of the spacecraft.

New proposals for experiments, some modified from the original Halley probe, arrived from all over the world, many of them from the United States. There was room and power in the spacecraft, however, for only 10. Among them were ultraviolet and infrared spectrometers, sensors to register mass and composition of the comet dust and the ions, and magnetometers to register electric charges.

ESA's Scientific Program Committee was now convinced that

a camera was necessary. A Giotto camera would give detailed color photographs of the nucleus, providing its true size, shape, and color.

The camera for Giotto had to be more complex than the one Delamere, Reitsema, and the others had originally designed for the canceled probe. In fact, their Giotto camera was (and is) perhaps the most sophisticated camera ever built. Their new designs solved many new problems, one of them having to do with Giotto's much higher encounter speed. According to the camera designers, photographing Halley's nucleus during the high-speed flyby would be like photographing the eye of a jet fighter pilot passing by at 2000 miles per hour.

The mission called for Giotto to rendezvous with the comet in March 1986 when it was closest to the Earth. March was a month after the comet's perihelion, after it had circled the Sun and was on its way back to beyond Pluto. However, the comet would still be extremely close to the Sun. The Sun is trillions of times brighter than Comet Halley, so that in addition to taking into account the speed of the encounter, Reitsema and Delamere had to design a baffle system for the camera that would block out the Sun but still allow it to photograph the (comparatively) much duller comet.

They also had to figure out a way to minimize the effect of the comet's dust particles and to compensate for the rotation of the spacecraft. Whatever designs they submitted had to be approved by Uwe Keller, a scientist at the Max Planck Institute for Aeronomy at Lindau, West Germany. ESA had selected Keller as the principal investigator for the camera, and he had already started to raise the money for it.

An independent Japanese mission to Halley, that nation's first venture into deep space, had already been approved by Japan's Prime Minister when Kunio Hirao came to the meeting of ESA's Special Program Committee in Paris.

Japan's Institute of Space and Astronautical Science (ISAS) had risen in status. Formerly a research group within the University of Tokyo, ISAS became a government agency under the

Ministry of Education, Science, and Culture. Actually the institutional ties with Japanese universities weren't completely severed, however, since ISAS was subdivided into nine divisions directed and managed by university professors who supervised graduate students in various research disciplines.

ISAS built its own new graduate campus at Sagamihara, a town outside of Tokyo that housed up-to-the minute research, development, and testing facilities. Institutionally there is nothing quite like this organization in the United States. We might have something like it if NASA, Massachusetts Institute of Technology, and California Institute of Technology were merged and then built their own, brand new graduate campus. To parallel the Japanese experience the Office of Management and Budget would have to say to this space and educational conglomerate, "OK, here's your money. You people know better how to allocate it than we do. Go do it."

The scientists and engineers of ISAS designed two identical spacecraft, small in comparison to Giotto: cylinders 4 feet 8 inches in diameter, 2 feet 4 inches high, and like Giotto, encased in solar panels. Lightweight batteries would power two or three experiments.

The Japanese craft were not designed to dive into the coma, as was Giotto. Instead they would skim by the comet at a distance of approximately 120,000 miles. The on-board experiments would analyze the comet's enormous hydrogen cloud, the hydrogen corona. A Skylab photograph of Kohoutek's corona, showing it to be larger than the Sun, had been proudly displayed to the conferees at Goddard by Whipple.

One instrument on the Japanese spacecraft would report on the activity of the hydrogen in the coma and the other on the interaction of the solar wind with the comet.

By 1980 ISAS had been launching a series of increasingly complex scientific satellites for over 10 years. At first ISAS depended on American rockets and such launch facilities as White Sands, New Mexico, and Wallops Island to deliver their satellites into orbit. Then ISAS developed its own launch site, the Kagoshima Space Center on the southernmost island of Japan, the sedge-covered rolling hills of Ohsumi Peninsula. ISAS flattened some of the hills and over an area of several square miles constructed a telemetry and tracking center, assembly buildings, a launch tower, an administrative center, and a conference center.

Kagoshima has a beautiful view of the Pacific Ocean and the beaches of a curved inlet. But the Ohsumi Peninsula is a fishing center for tuna. The fishermen watched apprehensively as the hills were shaved down. Red and white striped towers and new shiny white buildings appeared above the low woods, and the fishermen quickly discovered the twentieth century (if not the twenty-first) had arrived on the hilltop just outside their 500-year-old villages.

"*Hai*, rockets," the fishermen agreed. They had seen holiday rockets shoot up into the sky and explode into shimmering flowers, fountains, and dragons. But rockets rising from the now flattened hilltop would roar louder than the largest firework dragon and scare their tuna away.

These medieval fishermen may seem a little anachronistic in the age of deep space exploration and computer printouts, but they had political power and they knew how to exercise it. Fishing is an important industry in Japan. The fishermen had a strong union, headquartered in Tokyo and equipped with computer printouts and economic clout.

The same ministry that handles trade in Japan also directs the National Space Development Agency (NASDA), which launches all commercial satellites and would share the Kagoshima Space Center with ISAS. There was a great deal of polite discussion about what might be done about the rockets and the fishermen. First NASDA agreed to pay about $8 million down as compensation, then $4 million annually. In addition, all launches were ordered to take place only during the months of January and February or August and September, when the fishing industry was most slack.

The restriction concentrated launch schedules, but fortunately ISAS had already developed its new Mu rocket, the 3SII, weighing 70 tons, with a booster capable of launching 1700 pounds. Actually the decision to undertake a deep space mission started with a desire to give the Mu 3SII a practical test.

In addition to Japanese and European plans, the Russians were planning go to Halley, too.

Roald Sagdeyev, director of the Soviet Space Research Institute (IKI), is well known among American astrophysicists for his work in planetary spectroscopy. Conversant with the American space scene, Sagdeyev speaks English with something of the sentence rhythms of Franklin Roosevelt though with the Russian accent that Peter Ustinov seems to imitate. His professorial appearance is belied by an easy congeniality. A superb manager with Gorbachev's aptitude for public relations, he is the only "hero" of the Soviet space program who is not a cosmonaut. He is known popularly as Mr. Space, Comrade Cosmos.

When Sagdeyev attended the meeting of ESA's Scientific Program Committee, the Soviet Academy of Sciences, the governing body of the IKI, already had a plan for a simple flyby to Comet Halley. Sagdeyev, however, decided to push for a mission a great deal more valuable scientifically than a mere flyby.

For 20 years the Soviets had been practically commuting to the planet Venus with a spacecraft they called Venera. Veneras 9 and 10 had released landers that, parachuting through Venus' sulfuric acid atmosphere, reached the surface of the planet. From there they transmitted to Earth the first photographs of the rocky surface before they were destroyed by the atmospheric pressure. Veneras 11 and 12 also dropped balloons through the Venus atmosphere, which signaled more detailed information about the composition of the atmospheric gases.

Veneras 13 and 14 were already under construction; they were being designed to land softly and send back pictures of the landscape from two different locations on the planet. One of the Veneras, Venera 13, would drill into the Venusian rock, take in samples, analyze them, and transmit the results of the analysis back to Earth. The Soviet spacecraft had always visited Venus in pairs, the ultimate in redundancy. Veneras 15 and 16 were already in the design phase with instruments to be improved over those of 13 and 14 when Sagdeyev went to work lobbying for a Soviet thrust at Halley.

When Sagdeyev returned to Moscow from Paris, he and V. M. Kovetunenko, director of the Venera program, agreed to combine a Venera mission with a Halley mission that would be as impressive as it would be scientific. After dropping their probes into Venus, the two spacecraft would use the gravity of the planets to swing them toward Comet Halley, a trip of about 85 million miles. There the spacecraft would penetrate right into the atmosphere of the comet—a few days before the arrival of ESA's Giotto. The mission would score another first for the Soviet space program, always a powerful motivation for the Soviet Academy of Sciences. And the spacecraft would have a new name, Vega, an acronym of Venus, and Halley (*Galley* in the Cyrillic alphabet).

The mission would not be a strictly Soviet show. The Academy of Sciences agreed to allow international cooperation for the Halley mission. As ESA did, IKI requested proposals from other scientific bodies of the world for instruments that the Vegas might carry into the atmosphere of Halley. Sagdeyev also promised to make available to all the other Halley missions flying to the comet any information the Vegas gathered. This was glasnost before Gorbachev even announced it.

International cooperation for this fourth predicted return of Comet Halley began with a conference organized by the International Halley Watch (IHW) at Patras, Greece, in August 1982. More than 500 astronomers from 53 nations met on the shores of the mirror-blue bay near the base of Mt. Parnassus, home of the ancient Greek muse of astronomy, Urania.

The aim of the IHW was to coordinate all the worldwide observations of Comet Halley, whether on the ground or in space and provide a library where the results of all investigations would be preserved, cataloged, cross-referenced, and computerized.

Two astronomers looking at the comet at the same instant but in two different locations could report observing two very different phenomena, something in the ion tail the other observer wasn't watching for, a bright spot in the coma, or the formation of an antitail. In a way, multiple astronomers reporting on a comet demonstrate the truth of the old Indian story of the five blind men reporting on the different parts of the elephant they touched. And the comet, unlike the elephant, is an object that changes as it glides through the sky. The solar radiation and solar wind throw the dirt and the snow off the dirty snowball irregularly, in jerks and spurts. The same giant telescope peering out of the same giant dome an hour later might pick up something entirely different in its second view than it had during the first. The IHW would provide a central reference point for all the varying data, a day-by-day report, even an hour-by-hour report where necessary, a complete record of Comet Halley's visit.

Louis Friedman, a JPL scientist who had the original idea, wanted the IHW to be truly international. At Patras Ray Newburn of JPL and Jurgen Rahe of the University of Erlanger-Nurnberg, West Germany, divided the world between the Eastern Hemisphere—with headquarters in Bamberg, Bavaria—and the Western—with headquarters at JPL in Pasadena, California.

Astronomers were divided between observers and scientists. The observers were coordinated into an Earth-girdling network of the world's finest telescopes.

John Brandt, a leading cometary astronomer at Goddard, grouped the scientists in lists according to their speciality: those who had concentrated on the comet as a whole and those who had made a study of its various parts: the nucleus, the coma, the tails (all this accomplished by various techniques such as spectroscopy, radiometry, and photometry). With this organization, the IHW would provide guidance for and communication with astronomers so that Halley would be observed from as many locations as possible, at all possible times, and with all possible techniques.

Most importantly, the observers and scientists would send their data to the IHW archives at Pasadena for the first comprehensive library dealing with the appearance of a comet in the solar system. The response was enthusiastic. The IHW received the cooperation of over 1000 world-renowned astronomers from 51 different countries.

Steve Edberg of JPL sent word out that the IHW wanted to hear not only from professional astronomers but also from serious, organized amateurs. In astronomy *amateur* is not a derogatory term, but is used in its original sense of "lover." These amateurs are lovers of the night sky. There are more lovers of the night sky than there are jobs in professional astronomy. During the day these amateurs have to earn their livings as lawyers, accountants, auto mechanics, etc.

In the last decades, amateurs have built telescopes that would have delighted the professionals during Halley's 1910 visit. A moderately priced 35-millimeter camera of the 1980s would have exhilarated the 1910 astrophotographer almost as much as finding a personal comet. Modern fast films were beyond the 1910 astrophotographer's wildest dreams. Amateurs have equipped their telescopes with motor drives that compensate for the motion of the Earth. Some wealthy ones have even equipped their telescopes with electronic cameras that transfer the light from an object directly to a computer (CCDs).

The professional astronomer also loves the stars, but he or she spends as much time staring at a computer monitor as looking at the heavens. The amateur can stare into the face of his or her love all night long. These were the people Edberg wanted to hear from, and they numbered in the thousands.

IHW would produce an effective, worldwide, round-the-clock monitoring effort. This had never been done before.

After the conference at Patras, representatives from NASA, ESA, ISAS, and the Soviets met at the Kagoshima Space Center in Japan. All the nations involved agreed to form an interagency consultative group to exchange information. In that way, each probe would help contribute to the success of the one coming after. NASA took on the responsibility of providing tracking facilities all around the globe from JPL in Pasadena, California, to Australia, to Spain and back again, the best the United States could do thus far deprived of a Halley mission.

Far away from the scientists and engineers on Earth, Comet Halley continued in the elliptical orbit originally described by Edmond Halley in Islington, England, 300 years ago. The comet was sailing sunward between the orbits and Uranus and Saturn, still invisible to astronomers on Earth. Every large observatory was trying to be the first to welcome this celestial visitor back again.

6

Welcome Back, Comet Halley

At the McDonald Observatory in the Davis Mountains of deep southwest Texas, at the Hale Observatory in the San Jacinto Mountains of southern California, at the Special Astrophysical Observatory (SAO) in the Soviet Caucasus Mountains, and at many other observatories all over the world the largest telescopes were fixed on a specific point in the sky, a point along the orbit of Comet Halley. As yet the comet was barely a dim speck beyond Saturn, a black glob, a peanut shape about 6 miles across. This, the nucleus of Comet Halley, reflected only the distant light of the Sun toward which it was hurtling. This dot of reflected light still had to travel over 1 billion miles before telescopes on Earth could pick it up. Astronomers all over the world were racing to register the light of this dot, in a competition to be first announced to the world, "Halley is back!"

It takes a telescopic Gargantua to be first, and the gargantuan observatories are all on the tops of mountains: the Canada-France-Hawaii Observatory on Mauna Kea at 14,000 feet; the

University of Texas' McDonald Observatory on Mt. Locke at 6700 feet; the Hale on Mt. Palomar at 5700 feet; the Mayall on Kitt Peak, Arizona, at 7000 feet; and the Special Astrophysical Observatory on Mt. Pastukov at 6400 feet.

Telescopes are not on the tops of mountains to be nearer the stars but because the air is drier and clearer on the tops of mountains. The night sky seen from these mountains is both stupefying and intimidating. Endless numbers of stars glitter against the black night. You see the night as you've never seen it before, stars so close together they form veils flung across the heaven. Looking at this you realize there are worlds beyond worlds beyond worlds across the depths of space.

Here, beneath these amazing skies, humankind is confounded and needs a giant, a Hercules, for sustenance, something to match the infinite vault overhead. Surrounded by this immensity early astronomers dreamed of prodigious telescopes.

At the end of the eighteenth century, the English astronomer William Herschel constructed a telescope with a mirror of metal alloy that was 48 inches in diameter and weighed 1 ton. The tube of the telescope was 40 feet long and was suspended in a triangular scaffolding 50 feet high. For its time, the construction was a marvel of complexity, of size and, yes, unwieldy even for Herschel and his sister, Caroline, an astronomer and comet finder in her own right.

Then in 1848 William Parsons, Lord Rosse of Birr Castle, Ireland, viewed the sky with a 72-inch mirror. Rosse's telescope weighed 4 tons and was mounted between two brick walls 56 feet high. The telescope was called, appropriately, "The Leviathan," and nothing challenged it until the American George Ellery Hale hauled a 100-inch mirror on a grinding, chain-drive Mack truck up Mt. Wilson in the San Gabriel mountains of southern California. This was the start of the American century in the creation of telescopes. Hale himself surpassed his 100-incher with the 200-incher on Mt. Palomar completed in 1949 and posthumously named in his honor.

Palomar's 200-inch mirror is made of Pyrex, the same material as baking dishes and for the same reason—resistance to heat and cold, the heat of the Sun by day, the cold of the Mt. Palomar nights. The great mirror, covered with reflecting aluminum, is at the base of a rigid framework that looks like a section of a captured suspension bridge. Light passes through the frame to the mirror.

Floating on oil, the whole 530-ton construction follows the stars and moves by a tiny, 2-horsepower motor, now directed by a control board on the darkened observatory floor. Awed visitors in a glass-enclosed balcony can watch the giant structure turn silently and be hypnotized by its colossal magic, by its sublimity. The colossus seems capable of bringing to Earth the radiance of heaven's mysteries on its own, the tiny person at the controls beneath it having nothing to do with its power.

The only mirror in the world surpassing the Hale is at the Soviet Astrophysical Observatory on Mt. Pastukov, which gathers its light with a mirror 236 inches wide (19 feet 7 inches). On Kitt Peak, the Mayall telescope has a mirror of 158 inches; on Mt. Locke, the McDonald, 107 inches. As always there are smaller observatories distributed around the larger one—10 on Kitt Peak, four on Mt. Locke.

The mountaintop of a large observatory complex is a veritable village, housing the director, operators, technicians, accountants, and secretaries. There are dormitories for visiting astronomers, a computer center, a visitors' center, libraries, offices, machine shops, warehouses, and more. These communities are organized to bring comets, stars, and galaxies onto photographic plates, electronic detectors, computer monitors, and television screens.

At the McDonald Observatory some astronomers are always sleeping. The rooms in the dormitory [Temporary Quarters (TQ), and one of the best equipped in the world] are soundproofed and can be darkened with black shades. The corridors have signs, "Beware. People sleeping."

An astronomer working on one of the telescopes typically sets up around 3 P.M., booting up the computer, testing out the spectrometer, feeding the coordinates of some celestial object into the controls, and checking the weather, always checking the satellite map on TV for good "seeing."

The "run," as astronomers call the time they spend on the telescope, begins at the end of twilight in the control room away from the open dome, where they follow whatever they are studying on a TV screen as the Earth moves. The run is over when the sky changes, when the object the astronomer is following sets, or when the Moon rises and glares it out or if a weather front comes in. The success of an astronomical observation still has a lot of luck connected with it.

McDonald serves four meals a day, breakfast, lunch, dinner, and a midnight lunch for the observers. Over mashed potatoes, pork chops, steaks, limas, and tomato salad there's talk, griping about the weather, enthusing, and gossiping. Because there are fewer than 1000 full-time astronomical observers in North America, most of them university professors, members of the midnight crowd usually know each other, if not personally at least by reputation through papers and conferences.

All of them wanted to be the first to record the sight of Comet Halley as it approached the Sun. This "recovery" was really a balancing act between careful astronomy and academic prestige. No astronomer wanted to admit being influenced by the public attention Comet Halley was getting, even in 1981. Publicity is a nonscientific phenomenon. Also an astronomer should never mistakenly record some other miniscule object in the heavens and announce it was Comet Halley. Astronomers knew their miniscule object, of course, wouldn't get past the scrutiny of Brian Marsden at the International Astronomical Union's Clearing House, but Marsden's careful expertise wouldn't prevent the mistaken announcer from being the butt of jokes. Still it was important to be first with the recovery because the academic prestige that would ensue could translate into scientific grants.

The run at the Comet actually started in December 1981. That date was as early as the calculations showed the comet could—just maybe—be seen. It would have been the earliest the comet had ever been recovered, over five years before perihelion. On Christmas morning 1758, Johann Georg Paltizsch was the first to pick up the comet's return—which had been predicted by Edmond Halley himself. His sighting was three months before the March 13 perihelion.

The next time around, Father M. Dumouchel and astronomers at the observatory of the Collegio Romano in Rome sighted the comet August 6, 1835, three months before its November 16 perihelion. In Halley's first return in the twentieth century, Max Wolf made the first photographic-plate recovery ever at Heidelberg September 11, 1909, six months before April 20, 1910, perhihelion.

On its second return in the twentieth century, the attempt to pick up Comet Halley *five years* before its perihelion, February 9, 1986, was a challenge for everyone involved, astronomers, mathematicians, and technicians. It was a test of skill, patience, knowledge, and, as usual, luck. The astronomers began their quest with an accurate timetable of where the comet should be on any given day—its *ephemeris*, to use the Greek word astronomers have adopted.

The return of Comet Halley to its position closest to the Sun (the perihelion) has been predicted four times since Edmond Halley predicted its return for 1758–1759. The calculations are complex. The planets pull against one another in a tug of war and, as in a tug of war, the forces exerted depend on their size and mass. Unlike in a tug of war planets pull without a rope and from a distance because of their gravitational fields.

To predict the exact path of a cometary orbit the astronomer-mathematicians must know each planet's mass and figure how close the comet was as it flew by. The comet itself throws in its own complications since the Sun's radiation and wind cause irregular jets of ions and dust particles to erupt from the comet. These act like a satellite's propulsion system and slow, advance, or joggle the comet along its orbit.

Until the return of Comet Halley in the 1980s, these complex calculations were done by human computers using pen and ink. This involved a truly horrendous amount of work. After Edmond Halley, the first of these "computers"—and they called her just that—was Nicole de la Briere Lepaute, who, with two male associates, sharpened up Halley's prediction for the 1758 return. (The 34-year-old Lepaute was married to a famous clockmaker for whose book on clocks she wrote the chapter on pendulums.)

December 1758 and the comet were both approaching with great speed. Lepaute and her team frantically scribbled their calculations from early morning, sometimes not stopping for meals, then continued into the dark, squinting at their figures by candlelight. Lepaute drove one of her associates into collapse, but she finished the prediction before Comet Halley was recovered. She missed the date of perihelion by less than one month out of the 912 months of the comet's total orbit.

For the calculations of the 1980s, JPL astrophysicist Donald K. Yeomans had computers that worked a million and more times faster than Lepaute. Yeomans and his team published the most detailed and accurate timetable ever calculated for Comet Halley, one that located the comet in its orbit minute by minute. To check what they did, they ran their figures back to Chinese records of the comet's 239 B.C. appearance, proving that astronomers trying to pick up Comet Halley for the first time during the 1980's could safely run Yeomans' projections into the computer that aimed their telescope.

But in December 1981, even the largest telescope wouldn't be able to record the image of the comet. The telescope would have to be equipped with an image intensifier. Photons of light coming from a distant object hit a charge-coupled device, (CCD), knocking electrons out of the silicon chips of the CCD. These electrons are amplified electrically and are transferred to the astronomer's computer and can be translated into graphics.

The brightness of an object in the sky is rated by its magnitude and calculated much like a golf score. The brightest objects are under par. The full Moon is −12. The bright star Sirius is −1.6, and the star Procyon is right on the border, +0.38. At the end of 1981 and in the first months of 1982 the nucleus of Halley, having not yet developed a coma or tail, may have been +27.

Brian Marsden believed the Halley report wouldn't be coming in to the Astronomical Union's Clearing House for a while. The comet's magnitude would have to brighten to at least +24, and he believed this would not happen for a year and a half. He predicted no recovery until summer 1983.

Astronomers took Marsden's prediction as a challenge. Michael Belton, Director of Kitt Peak, had scheduled time on the 158-inch Mayall. When the time arrived for the observing run, the telescope was pointed to coordinates entered in the computer near the star Procyon in Canis Minor. Belton hoped the equipment was strong enough to register an image. Comet Halley was probably still around +26.

The astronomers were not at the telescope itself but at computers in the control room. At best the comet would register only as a few dots on their computer screen! If the dots moved in the right direction, clockwise, and the correct distance for the time elapsed, a miniscule 4½ arc seconds per day, then the light of Comet Halley had reached Kitt Peak on Earth.

The same process, with minor technical variations, was taking place all over the world: on Mt. Pastukov, on Mt. Locke, at Mauna Kea, and at Cerro Tololo at the foot of the Chilean Andes. Even The Big Eye, the 200-inch Hale on Mt. Palomar was searching space.

But in December 1981 Marsden was proven right. Comet Halley was still too far away, its light too faint to produce energy enough to affect a silicon chip or create an image on a sensitized film. However, the comet was closing in on Earth, and the signal could only get stronger. Belton told his team at Kitt Peak to keep working. It was only a matter of time.

Some of the other observatories kept at it too. During January 1982 the comet was still too far away. February, however, presented a new problem: observers had to search through a field of asteroids circling between Mars and Jupiter. On a computer, asteroids register like troublesome points moving against the background.

However, February offered the last chance for a recovery until fall, because in the spring Halley disappeared behind the Sun. Halley would reemerge in October 1982 and then at magnitude +24.

Telescopes and CCDs went into action during February. Astronomers at the McDonald Observatory decided to try with their largest telescope, the 107-incher. And McDonald scientists are equally proud of their Griboval electrographic camera, designed by Paul Griboval. The camera covers a wider angle of the sky than most of the video intensifiers, certainly more than any CCD. The silicon chips of the camera are so sensitive that under the right conditions they can pick up a single photon of light.

The run at Comet Halley by the McDonald team was conducted in the glare of publicity, reported by Terry Dunkle in *Science 82*. The team, Ed Barker, Derral Mulholland, and Elizabeth Bozyan, fitted the 107-incher to the Griboval electrographic camera.

Their computer was able to follow the position of the comet according to Yeomans' calculations, The team made plates over three nights, exposures from a half-hour to 2 hours long, developing them almost as quickly as they were recorded, and printing them on 8-by-10-inch high-contrast paper. They found the right kind of fuzzy object, and it moved all right. Ed Barker took five plates back to the University of Texas at Austin and projected them on a 19-inch television screen. Unfortunately no fuzzy object moved at the rate it should have if it had been the Comet. The objects proved to be asteroids.

Harlan Smith, Director of McDonald, believed the Comet was still a +27 magnitude. He told his staff, "No one's going to get it for a while yet."

He was right. No one got it in February 1982. Astronomers

working on other projects sent abstracts of their papers to Larry Esposito, a Research Associate at the Laboratory for Atmospheric and Space Physics (LASP) in Boulder, Colorado. He was organizing the annual meeting of the Division of Planetary Sciences of the American Astronomical Society, where an exchange of the latest astronomical advances take place.

The conference was held at the Harvest House in Boulder, a hotel within walking distance of Folsom Field, home of the University of Colorado football team, the Buffalos. Fortunately the Golden Buffalos were playing in Kansas that weekend, October 16, and there was room at the inn for the astronomical pundits. Esposito received abstracts of over 200 papers and carefully scheduled their readings and discussions.

After the conference had been in progress for a couple of days Esposito received a telephone call from Edward Danielson at Cal Tech at Pasadena. After Danielson had submitted his ideas for the Halley-Tempel camera to Reitsema and Delamere, he and one of his graduate students, David Jewett, thought they would have a try at recovering Comet Halley.

They borrowed the best CCD they could obtain, the engineering test model, a duplicate of the one to be used in the Hubble space telescope, then being prepared for a shuttle launch scheduled for 1986. There wasn't a more sensitive CCD anywhere in the world.

Danielson had called Esposito to tell him they had recovered Comet Halley on the Hale telescope tied into that CCD, and Danielson asked if he and Jewett could be given time for a presentation during the conference.

Esposito is a quiet person whose reputation in astronomy comes from his discovery of a volcano on Venus. Keeping his excitement over the Halley recovery under control, Esposito calmly agreed to squeeze Danielson and Jewitt in among the papers. By the time he posted the change of schedule on a bulletin board, the news of the recovery had spread. Every astronomer at the Havest House knew Halley was back.

Anita Cochran, a specialist in cometary spectrography from

McDonald was on the phone in her room when her husband interrupted her with the news. A group of the conferees were drinking beer and eating sandwiches at the New York Deli, popular from the *Mork and Mindy* show on TV. A jogger ran the news to them and earned a free beer.

Esposito called the local papers. The Boulder *Camera* and the Denver *Post* beat *The New York Times* in announcing the recovery of Comet Halley for this visit to planet Earth, its 30th recorded visit, beginning with the Chinese sighting of May 239 B.C.

The difference between the attempt at McDonald and the success at Palomar was the difference of about 3.4 million miles the comet had moved closer to the Earth. Now it was brighter by nearly three magnitudes, at +24.2. And the luck of the weather made a difference too. Danielson had traded a later date he had reserved on the Hale with someone who had the predawn of October 16. And that night it was clear for 960 million miles.

In the early morning of October 16, 1982, long before dawn, Danielson and Jewitt had carried their borrowed CCD up an elevator at the side of the giant dome, across a walkway 50 feet above the floor, and into the prime focus cage. They adjusted the CCD case to the large, round pedestal in the center of the cage, at the concentration of light. Then the CCD was wired to a computer in the data room where Danielson and Jewitt returned to make their run. Outside on the floor, Juan Carrasco, the operator of the Big Eye, swung the machine to the slit in the dome open to the sky and to the coordinates Danielson and Jewitt gave him.

The two astronomers made seven 8-minute exposures between 10 and 15 minutes apart. For the next two days they checked and measured each dot on the seven frames. Unlike the dots recorded at Kitt Peak and McDonald, one of these dots moved the way Halley was supposed to, clockwise, and at the right speed.

Three nights later Danielson and Jewitt returned to the telescope again. The dot had moved near a bright star in the Milky Way which outshone it so the comet's light wasn't distinguishable. But at Harvard, Marsden accepted Danielson and Jewitt's October 16 sighting. Halley was the ninth comet of 1982 to be discovered or recovered and was designated 1982i by the International Astronomical Union. The names of Danielson and Jewitt would now join those of Johann Paltizsch, the Reverend Dumouchel, and Max Wolf.

Danielson and Jewitt found Halley in its orbit only 8 arcseconds (37,000 miles) off Don Yeomans' calculations. Not much had happened to Comet Halley during its 76-year solo journey that Yeomans had not fed into the computer.

While Halley was as yet only fuzz on computer printouts, the Tasco corporation, manufacturers of binoculars and other optical products, were looking three years ahead to 1985. They were preparing their Halley Comet-Watching Kit, consisting of binoculars, a poster showing the path of the comet, and a diary. The first of a T-shirt deluge arrived on the market with the legend announcing, "I'm like Comet Halley. A once in a lifetime experience."

A meeting of "The Halley's Comet Society of London" celebrated the recovery. The society is actually a select club whose members are not scientists but socially prominent English men and women none of whom had ever seen Comet Halley nor had plans to observe it during the comet's current appearance. The club does not seek new members nor publicity, and its celebration of the recovery would have passed unnoticed had not a correspondent of *The New Yorker* reported the meeting.

Champagne was served in silver goblets, and the founder of the Society, Brian Harpur, presented evidence that Edmond Halley's name was prounced "Haw-ley" since Queen Anne (1665–1714) spelled the name that way in a letter to the astronomer. Patrick Moore, author of *An Introduction to Halley's Comet*, pronouncing the name "Hal-ley," gave a slide lecture which concluded with a picture of the Danielson-Jewitt recovery. The meeting adjourned with the singing of the Halley ballad, "Glory, glory, Mr. Halley," to the melody of "The Battle Hymn of the Republic."

Robert Subranni, of Linwood, New Jersey, was working on the dies of a Halley medal with the face of Edmond Halley from late in his life on the obverse and the comet in the shape of a dragon on the reverse. It was to contain the motto: *Libera nos a malo cometatae, 1986* ("Save us from the evil of the comet"), a prayer from a litany preserved from the time of Shakespeare. The average Londoner at that time agreed with the opening lines of *Henry VI, Part One*:

Hung be the heavens with black, yield day, to night
Comets, importing change of times and states,
Brandish your crystal tresses in the sky.

7

Hot ICE

*

When NASA canceled the Halley Earth Return (HER), the last attempt by American astronomers for a Halley encounter, its disappointed champion, Bob Farquhar, didn't rush around pounding on doors for it. He was playing with an idea that originated with events that went back to 1978.

Farquhar had helped plan the third International Sun-Earth Explorer, ISEE-3, a spacecraft now in high Earth orbit that had been launched from Kennedy Space Center August 11, 1978. Ever since November 1978, ISEE-3 had been circling quietly 930,000 miles away between the Earth and the Sun, carrying ESA and American-sponsored instruments.

ISEE-3 orbited at a very special location in space, Libration 1. The root of the word, "Libration," is Latin *libra*, "balance." This point is a balance point, an equilibrium point, between two gravitation fields, that of the Sun and the Earth.

The existence of such balance points was first theorized by an eighteenth-century mathematical genius, Joseph-Louis Lagrange. The idea of balance points in space fascinated Farquhar when he was a student at Stanford. He wrote his doctoral dissertation on them, and led the first navigation team to try the experiment by positioning ISEE-3 at Libration 1.

Lagrange, of couse, had been right, and the balance held. ISEE-3 circled around that libration point for three years expending very little fuel as its 12 experiments monitored the composition of the Sun's radiation and its interaction with the Earth's magnetic field.

ISEE-3 was a small, 16-sided aluminum craft, slightly larger than Giotto, 5 feet across, and 6 feet high, weighing half a ton. As with Giotto and the Japanese spacecraft, solar cells covered two-thirds of its outer shell. ISEE-3 bristled with nine antennae of varying thicknesses and lengths, the longest 100 yards from tip to tip. In flight the craft was controlled by 12 hydrazine thrusters.

Farquhar had designed ISEE-3 with extra fuel with the usual precautionary NASA planning that the fuel might be necessary to correct launch errors. However, even during the planning phase, Farquhar was wondering if sometime in the future he might be able to move ISEE to another location for another mission.

The future was now, and Farquhar was still thinking U.S. Halley mission, in spite of the official letter that Andy Stofan sent to Bruce Murray at JPL in September 1981 to stop work on all Halley missions. Farquhar believed it would be possible to use ISEE-3's extra fuel along with the impetus from a lunar swing-by to guide ISEE in a trajectory across space to Comet Halley and then dive it through the coma.

There were good arguments for trying this Halley option. ISEE-3 was already up there. The expense of the launch was over; the instruments had survived the shake, rattle, and stress of ascent. Better still, seven of ISEE's instruments were well suited to study the comet. The ISEE-3 Halley mission would be cheap, cheap, cheap, perhaps $3 million, one-hundredth the cost of any other deep-space mission. Perhaps just cheap enough to interest the NASA administration and give the United States a shot at Halley after all.

Computer printouts showed the gravity assist from the Moon and the subsequent path of the craft would take it to the comet in March 1986 when Giotto, the Vegas, and the Japanese craft had also all gathered there. The comet, however, would still be about 96 million miles away from Earth, too far for the 5-watt signal from ISEE-3 to reach the tracking stations.

It was frustrating. During the planning phase of the ISEE mission, Farquhar had also designed a stronger signaling system. But the budgeteers cut down the too-vigorous signal.

Farquhar heard "Halley, no" for the second time. Comet Halley was The Comet, of course, but, it wasn't the only comet in town. Tempel 2 had been talked about, but it was eliminated for an ISEE rendezvous because Tempel 2 would be even further away than Halley. However, Comet Encke, once considered for an ESA mission, was a definite possibility.

There were several other possible comets as well, and data on these were run through the computers. Among them was Giacobini-Zinner, a tiny comet first discovered by Giacobini at the Nice Observatory in France in 1900. It moves in the same direction as the planets, the opposite of Halley's direction, in an orbit lasting about six and a half years.

It was so faint astronomers kept missing its returns. Zinner saw it for the second time in 1913 from Bamberg, Germany, and then the dim comet came in and out of the solar system without being spotted until 1926. It was last seen in 1979 at magnitude +20.5 by the Naval Observatory, Flagstaff, Arizona, and photographed by Elizabeth Roemer.

Farquhar had seen one of her photographs and became a Giacobini-Zinner (GZ) fan. Don Yeomans of JPL, who calculated the orbit of Halley so accurately, had figured GZ's orbit for his doctoral dissertation at the University of Maryland. John Brandt, one of the organizers of the Goddard conference and the author of a standard text on comets, together with Yeomans projected the orbit during GZ's 1984–1985 return. At certain observatories, however, astronomers worried that GZ would be so faint they might not be able to pick it up at all. In addition, there was a possibility that, about when ISEE would be heading for it, the comet might split into pieces as Comet West had done in 1976.

Farquhar ignored these possibilities. He figured out a possible trajectory for ISEE from Libration 1 to the projected position of GZ close to its perihelion, September 5, 1985. If ISEE survived the dive through GZ's coma, Farquhar thought he still might be able to sneak it into the Halley flotilla by sailing it across space to monitor the solar wind as it blew toward Comet Halley.

ISEE had no special shields. Would the dust particles tear up the solar panels? Would they rip off the antennae or bend them back against the hull of the craft so they could no longer function? At least ISEE would find out, and Giotto and the Vegas would know ahead of time what they were getting into.

When word of the possible new mission for ISEE got out, the American and European scientists who had the chief responsibility for the solar wind experiments on the spacecraft—the principal investigators, as they are called—set up an angry outcry. If the purpose of their experiments were changed, their grants might be cut off. The ISEE trip through Giacobini-Zinner was secondary, they complained. Farquhar just wanted to be first through a comet. He was merely trying to upstage the Russians, the Japanese, and the Europeans.

Farquhar threw the growling principal investigators a bone. The solar wind not only forms tails on comets but also reacts with the Earth's own magnetic field. The wind forms an electrically charged magnetosphere around the Earth with a tail streaming out behind, its magnetotail.

The path that Farquhar projected from the libration point to GZ sent ISEE on five trips between 30,000 and 900,000 miles distant on the side of the Earth away from the Sun and took the spacecraft through the magnetotail into a previously unexplored area. The instruments on board ISEE could monitor this area, which had never been monitored before. This new mission was endorsed by the National Academy of Sciences, the National Science Foundation, and even the Space Science Board and Astrophysics Division of NASA.

David Dunham, Farquhar's assistant at Goddard, began working out the details of the ISEE flight through the magnetotail around the Moon and across to the comet. Every object had to be just in the right place at the right time, the Earth, the Moon, and Giacobini-Zinner. ISEE had to approach each object at the right speed and the right angle and be lined up so it would leave each object correctly headed for the next. Each hydrazine jet had to be fired exactly at the right time to change the craft's direction and for long enough to produce the estimated velocity and direction of ISEE, taking in consideration at the same time the gravitational forces of the Earth and the Moon. What resulted was the most complicated celestial acrobatics ever figured. Plotted on paper, ISEE's tangled path to Giacobini-Zinner looked like the Gordian knot.

ISEE had to move out of the libration point by June 1982 if the Earth, Moon, and GZ were to be lined up for its flight path. Farquhar still had not received official approval for the move from NASA. Just a matter of time, but how do you reshuffle the Moon? Farquhar notified NASA in writing. "We move June 10, unless you advise us to the contrary." NASA didn't advise them to the contrary or otherwise, and ISEE was well on its way to the Earth's magnetotail and to GZ.

ISEE sailed through the tail, out again, was hauled back again by the Earth's gravity, out again, back again, on each trip making five whips around the Moon, sailing very close to the Moon to utilize the maximum effect of its gravitational field. On November 21, 1983, ISEE's jets were fired to line it up for its lunar slingshot journey to Giacobini-Zinner.

Three weeks later ISEE glided into the Moon's cold shadow, out of the warmth of the Sun. ISEE's batteries were inoperative from its days at libration. This presented a problem since its power came directly from the bands of solar cells around the shell of the spacecraft which, of course, wouldn't function in the shade.

Before ISEE went into the shade, all its heaters were turned up to leave the spacecraft with as much warmth as possible for the frigid shadow run. At Goddard, project managers could only wait out the 28 minutes, wondering if there would be permanent damage when the craft reemerged. Would it even respond to commands again?

ISEE emerged from the shade, and the Sun took a few tantalizing moments to thaw the craft out. It had suffered no damage. NASA announced that ISEE was to be renamed the International Comet Explorer (ICE) in honor of its new mission.

ICE's five whips around the Moon had increased its speed from 2908 miles per hour to 5145. This was December 22, 1983, and ICE was headed toward Giacobini-Zinner. In 21 months ICE would be the first man-made object to encounter a comet. The unscientific word "first" was now being shouted right out loud in all of NASA's facilities. But no one had seen Giacobini-Zinner for five years. Where ICE was headed existed only on a computer printout. However, Yeomans promised the comet would be there!

8

End Run to Halley Mission

The early eighties were years of a shrinking NASA budget, and ICE was an inventive idea that stretched these shrinking funds. Then almost as soon as *Columbia*, the first reusable space shuttle, was launched on April 12, 1981, funds were stretched by another inventive idea.

Under the direction of Len Arnowitz, the Special Payloads Division at the Goddard Space Flight Center offered NASA a reusable robot spacecraft, Spartan. Arnowitz is an expert space engineer whose career began back in the pioneer days of the 1950s and the Vanguard project. Arnowitz and the Special Payloads Division of NASA had designed Spartan to be loaded into a shuttle cargo bay, taken aloft, deployed overboard, picked up again, and returned to Earth. Spartan was equipped with battery power, navigational equipment, and a tape recorder to store information. Scientific instruments, which formerly were launched on rockets, could be loaded into a Spartan carrier and tied into the systems already aboard. The instruments and Spartan would ride the shuttle into space as a unit, there to be deployed where the instruments would perform their experiments. Afterward, Spartan and the instruments would be retrieved by the shuttle and returned to Earth where the instruments would be unloaded. The Spartan, like an airborne pickup truck, could then be reloaded with another set of instruments, and sent back to go on another shuttle ride.

Spartan was a cheap way to get science into space. NASA had a carrier that could be used again and again. Scientific laboratories needed to assemble only half a satellite, the instrument side of the experiment, then integrate that into the electronics of the carrier already prepared at Goddard. In addition, instead of having a mission that, launched on a rocket, flew in space only 5 minutes, they could have a mission that lasted for days. And for the same amount of money.

The Laboratory for Atmospheric and Space Physics at the University of Colorado at Boulder had a proposal for the Spartan project. Toward the beginning of 1982, Charles Barth, Director of the laboratory, and a group of LASP engineers were checking over the proposal. They were Sam Jones, LASP's supervising engineer, who was as good a psychologist as an engineer; Fred Wilshusen, LASP's chief rocket engineer; and Bob Leyner, who reconstructed antique planes for a hobby and was in charge of the mechanical operation of the mission.

Barth had launched rockets for JPL and then joined LASP in 1965. Moving LASP into the Mariner program, he had developed special spectrometers designed to scan the atmosphere above Venus and Mars. On November 13, 1971, after a six-month flight, Mariner 9 carried 150 pounds of sensitive instruments, including LASP's most sophisticated spectrometers, on two daily orbits over the most intense dust storm in the history of Mars. Early in 1972 when the dust finally settled, the instruments radioed the radio telescope at Goldstone in the Mojave Desert what they had seen and measured. Scanning the ultraviolet bands of the Martian atmosphere, the LASP spectrometers recorded the temperature and detected hydrogen for the first time.

The engineers at LASP had designed spectrometers to be able to look at a very weak emission of light in the presence of a very bright emission from great distances. The weak emission here was the upper atmosphere of Mars; the bright emission was the surface of the planet. Barth and the LASP team realized they could use the same techniques they used for Mars on comets. Comets, like planets, have an atmosphere, their coma. The LASP spectrometers could scan the coma of Comet Halley, a very weak light, in the presence of a a very bright light, the Sun, and closer to the Sun than even Skylab had scanned Kohoutek.

Barth was principal investigator (PI) for what LASP hoped would be the U.S. Halley mission. The PI is the quarterback, the commander in chief, of a scientific experiment. The PI has to be a scientist with a solid reputation. Barth was certainly that. With a Ph.D. from the University of California at Los Angeles, he had been PI on so many experiments that he'd forgotten all but the most exciting ones like the Mariner missions, the previous Moon flight, Apollo 17 (which left one of LASP's spectrometers on the Moon's surface), and the Solar Mesosphere Experiment, the first wholly university designed and constructed communication satellite.

It is the PI who has to get the grant money for the experiment. Using the LASP spectrometers as a selling point, Barth sent a proposal for a Spartan-Halley mission to NASA headquarters in 1982 and requested a budget of $700,000, because he knew that was all NASA allocated for each Spartan mission.

In 1983 LASP got the approval and later the money for the mission, numbered Spartan-102/Halley. Stofan's "Stop-All-Halley-missions" letter, which had been sent to Murray two years earlier, never even came up. By the time NASA gave LASP the OK, the Spartan had been so popular that other experiments had been approved earlier. Spartan-102/Halley was number eight and scheduled to go into space about 1992.

Barth hurried to Washington to remind NASA that by 1992 Comet Halley would be beyond the orbit of Saturn heading toward Uranus. So NASA advanced Spartan-Halley to fourth place as part of the payload of the new shuttle, *Discovery*. The flight would put Spartan-Halley up in space between January 8 and March 7, 1986. This was the time span, the window they called it, that LASP needed. It was the period when Comet Halley would be most active, throwing out jets of dust and gas like an airborne Mount St. Helens.

Their new deadline gave LASP only three years to design instruments, construct them, test them separately and together, then integrate their observatory package with the Spartan carrier at Goddard. LASP had worked with Goddard on rocket missions many times before, and Barth was confident LASP and Goddard together could make that space window in time.

Barth estimated the $700,000 still wouldn't be enough to put the project together and that he would have to allocate funds from other projects. LASP received $225,000 each year to fly rockets and was committed to shoot off one more at the beginning of 1984. Barth believed that the Spartan-Halley budget was important enough that he could juggle three years worth of future rocket money into it.

Fred Wilshusen, LASP's chief rocket engineer, didn't object. He had helped Barth on the Spartan-Halley proposal and was enthusiastic about the mission. Raised on a Colorado ranch, like all ranchers he had had to learn to fix anything, hay rakes, Fordson tractors, and deep-well pumps. Like Pete Conrad on Skylab, he could still fix anything, though unlike Conrad, he hadn't had to fix things while they were in space. He was a graduate from the University of Colorado in electrical engineering. Early a co-director of LASP, he was a rocket pioneer with 35 years rocket experience going back to the days of Wernher Von Braun and the early rockets: the Nike-Cajuns, the

Aerobees, and continuing to the Orions that LASP was using in 1984. Wilshusen would have no difficulty adapting his renowned rocketry skills to Spartan and a shuttle launch.

LASP now had the engineers, the spectrometers, the instrument makers, and the facilities to construct this U.S. Halley mission. In addition, Barth needed a project scientist, someone to handle the day-to-day supervision. The project scientist needed to be conversant with cometary astronomy, engineering design, and theory and would have to work closely with engineers and other scientists, making sure the spacecraft and its instruments were designed to ensure the mission's success.

Spartan-Halley posed a unique problem for the project scientist. Once deployed from the shuttle, it was a robot, on its own. It had to fix on a moving comet, invisible in the glare of the Sun. However, no one would know the exact date or time when Spartan-Halley would be deployed into space, nor its exact position in relation to Comet Halley, until after the shuttle was launched. But somehow a way had to be figured out of entering all that information into the Spartan computers, its microprocessors.

Alan Stern was working at Martin Marietta in Denver when Barth interviewed him for the job. While Stern was only 26 and looked even younger than that, he had already accomplished a great deal. He had assisted with the design of the Manned Maneuvering Unit, a jet-powered chair that astronauts use as a personal spacecraft when they leave the shuttle and move around in space on their own.

Stern had four degrees from the University of Texas, Austin: two bachelors, one in physics and one in astronomy; two masters, one in aerospace engineering and one in planetary atmospheres. He earned the money for his tuition working as an intern in mission control at Johnson Space Center. For Stern, Johnson was the home of his space world. Barth hired him not only because of his accomplishments but also because Stern knew shuttle procedure well and no one else at LASP really did. For all his scientific background, Stern's knowledge was primarily theoretical, and that was also just what the project scientist needed.

LASP was already committed to one last experiment to be launched on a rocket, a Taurus-Orion. Riding the two-stage rocket would be the Ultraviolet Aurora 1984 (UVA 84), traveling 115 miles above the Earth into an aurora borealis.

While the LASP engineers worked on preliminary Spartan designs, Barth assigned Stern to a team that would test UVA 84 at Wallops Island Test Station, just off Cape Charles Peninsula, on the east coast of Virginia, then launch it from Poker Flat Rocket Range, in Alaska, 20 miles north of Fairbanks, and 4400 miles north of Johnson Space Center, as a rocket flies.

Wilshusen was in charge of the UVA 84 team, and there were enough similarities between this small satellite and the Spartan-Halley hardware to provide break-in experience for both Stern and the third member of the rocket team, Rick Kohnert. Kohnert had started working full-time with LASP only seven months before, though he had worked there part-time while he was still an engineering student at Colorado specializing in optics. Wilshusen quickly discovered Kohnert was a fast learner and skillful with circuitry. Kohnert seemed to understand how machines worked intuitively, as if he were part of them.

At Wallops Island the spectrometer and other hardware of the satellite were tested in concert with the NASA telemetry and guidance systems that would ride with them into space. Then the LASP team flew up to Fairbanks and Poker Flat for the final assembling and the launch.

Poker Flat is not one of the better known NASA installations. Operated jointly by NASA and the Geophysical Institute of the University of Alaska, the range had been bulldozed out of the tundra and pine of the Chatanika Valley where the Poker River runs. An early range manager, a student of American literature, borrowed the name from Bret Harte's famous short story.

Since 1960 Poker Flat has been a special center for studying the atmosphere and the aurora borealis. More than 200 rockets have been blasted as high as 1000 miles above the Flat since 1960. However, the range still seems to be a well kept secret, even in Fairbanks.

When Wilshusen, Stern, and Kohnert arrived, most of the visiting NASA personnel occupied one floor of a downtown hotel. Here the LASP crew met John Ansley. Ansley was an outcast at Poker Flat from some NASA center in the lower 48 for over a year, lent as a technician at the range.

Ansley had scouted all the bars and all the women. He always tried to impress them with the glamour of space and described how he punched satellites right up into the Northern Lights. A brunette at one of the bars didn't believe him. She had it on the best authority that no rocket would survive in the Northern Lights—the aurora borealis. Ansley took the easy way out. He lied.

"Sorry about that line," he said. "I'm really a tundra picker."

"A tundra picker? What's that?"

"We pick tundra," he told her. "We pack it and ship it back to the states."

"What do they ever do with it?"

"It keeps stuff fresh like fruit and vegetables. Fresh tundra for fresher produce." The story was so bizarre that it was believed. From then on John Ansley was "Honest John, the tundra picker."

But how was John to explain the aurora borealis? If he said an aurora is actually nature's television spectacular, would he have been believed? That sounds as crazy as the tundra pickers, so maybe he would have been believed. It's true, actually. The Sun is the broadcasting studio. It shoots out particles, electrons and protons, from its surface. The Earth's magnetic field forms the picture tube. Although most of the Sun's particles sweep by the Earth, some drift down into the polar regions and produce the show, the swirls, bands, weaving curtains in blue, green, white, and red on a celestial screen 600 miles long. Only those planets that have a magnetic field can tune into the Sun's programs—others being Jupiter, Saturn, and Uranus.

That stream of electrons and protons shooting off the Sun is the solar wind, whose existence was first confirmed by Explorer I in 1957 and which was the phenomenon the experiments of ISEE-3 were monitoring when Farquhar diverted the spacecraft (now renamed ICE) toward Giacobini-Zinner. The wind sweeps from the Sun at between 180 and 300 miles per second (or between 64,800 and 108,000 miles per hour). In 1716 Edmond Halley was the first person to suggest the aurora might be related to the Earth's magnetic field. But Halley didn't know anything about the solar wind, that it is the same force the creates the aurora and also makes his and other comets into shimmering globes with long tails.

Of all the instruments in the UVA 84 payload that would be registering the interaction of the solar wind with the Earth's magnetic field, Wilshusen and the LASP team were most concerned about the spectrometer. To analyze the ultraviolet bands of the aurora, it had to operate in a vacuum. Actually engineers can cheat. They can use nitrogen instead of a vacuum because nitrogen is transparent to ultraviolet light.

Wilshusen was provided with three tanks of nitrogen to pump into the spectrometer and drive out the Poker Flat air. Before he started pumping, Wilshusen took the hose on the tank, turned the valve, and sniffed the nitrogen. He didn't change his expression. He turned the gas off, then on again and sniffed a second time. Then he turned the valve completely off, stood up, and walked outside.

Stern watched him. In Alaska you don't walk outside for no reason, especially without putting on a hat, coat, and gloves. Stern was about to go outside when Wilshusen came back in, went over to the tank, and sniffed the gas a third time.

"Alan, come over here," he said. "Smell this." Stern did and picked up the faint odor of oil. "Is this the nitrogen or is it me?" Wilshusen asked.

"Something's in there," Stern answered. "That's oil." Wilshusen called Kohnert who was off testing the photometers. Kohnert came over, and he, too, smelled the nitrogen.

"There's oil in that stuff," Kohnert said.

Wilshusen hadn't announced, "I smell oil." First he had gone outside and stood in the cold to make sure he had a perfectly clear nose. Then he came back and sniffed again. He still didn't shout. Smells are very suggestive. Wilshusen wanted Stern and Kohnert to make up their own minds, so he just asked them to take a whiff. Oil is made up of hydrocarbon molecules that soak up ultraviolet light. If there is oil in a spectrometer, no one will ever receive any data from the instrument.

"This is the wrong damn nitrogen," said Wilshusen. The LASP procurement department had ordered water-pumped nitogen. However, someone put the nitrogen into the tank using pumps lubricated with oil, and somehow oil leaked into the nitrogen.

"What are we going to do way up here?" Stern asked. There was no certified, water-pumped nitrogen in all of Fairbanks or Anchorage. They could have LASP send it up. It meant at least a two-day delay, but delay it had to be.

To Stern and Kohnert, Wilshusen had just demonstrated the mentality of people who have been in the business of space for a long time. In space you can't ever take a chance. You must take every precaution. Sure, if the oil had ruined the experiment, Wilshusen could have blamed the supply company: they had certified contaminated nitrogen. But it still would have been Wilshusen's fault. Barth would have asked him, "Did you check the nitrogen, Fred?" At Poker Flat for his 71st rocket, Wilshusen didn't have to be reminded.

Wilshusen still had to create a vacuum in the detector, a device in the satellite that would transmit information collected by the spectrometer to the telemetry room at Poker Flat. Wilshusen couldn't cheat with nitrogen this time. However engineers, with a boost from physicists, have a slick method for achieving a high-quality vacuum in the detector. They use an ion pump. The ion pump, true to its name, creates ions, like those in the ion tail of a comet. The pump sends an electric charge through the signaling device, and rips off one electron, creating a positively charged ion. Now the ion can be trapped magnetically and sucked out of the signaling device, leaving a vacuum inside.

There would be an ion pump on Spartan-Halley, and Stern was learning about the necessary care on a rocket. In addition he was learning to withstand cold. Astronauts, he knew, received special training against the −20 degree Fahrenheit cold. His own training consisted of jogging a mile every day from the Assembly Building at Poker Flat to Chatanika Lodge, where the team ate lunch.

The warmest outdoor object in Chatanika Valley was the Taurus-Orion rocket with its satellite. When UVA 84 was integrated with the rocket, the unit was hauled from the assembly building to the launch tower. Here it was wrapped in styrofoam strips glued together, and hot air was pumped into it until the crew received the order to launch.

Back in Boulder, Barth was following auroras over Siberia by satellite relay, hoping to spot one worth a launch 24 hours before it arrived over Alaska. He knew the best aurora in 1984 would probably occur between February 23 and 28.

Scientists and engineers are not just human circuits and capacitators, nor does their manipulation of electrons prevent them from appreciating the beauties of nature. The LASP team never tired of watching auroras. Amost as soon as the haze of the night Sun vanished, the first streamer of the aurora began, white flickering in the blue night above the vermilion band on the horizon. The flicker expanded slowly into vertical streaks, a shimmering glitter that swayed with an awesome dignity.

No two nights were the same. Sometimes the aurora formed parallel bands of a purple lighter than the sky. Each band converged into a point, somewhat as if it had been left there as the vapor trail of a giant, invisible, spacecraft.

Other auroras formed separate parallel streaks of blue, and through them Stern and Kohnert could see the stars. Sometimes a yellow curtain looked like thin, but large, God-size icicles, alive and shimmering in a slow sway.

The night of February 26 and the morning of February 27 above Poker Flat the aurora was especially dazzling, a triple curtain of light, rolling, swaying slightly. The snow below reflected the pink with a faint tint of blue. Barth made his call and it was "Go."

A 4-hour countdown began. Blockhouse stations were staffed. Steve Bonham, the range safety officer, notified airfields, civilian and military, that a rocket would be in the air at 1700 UT, universal time, Greenwich mean time, 9 A.M. Pacific mountain time. He radioed NORAD (North American Air Defense) in Colorado Springs, and NORAD relayed the message to the Soviet defense network so Soviet radar and infrared sensors didn't mistake the flying Taurus-Orion for the start of Armageddon.

Wilshusen was responsible for the instruments of the experiment. He had gone through this routine 70 times before, but he didn't act as if he had, but rather as if it were the first and only time. In a very real sense this was true. The flight is the only time, the one time the rocket rides, the one time the instruments see and tell.

Wilshusen watched all the needles very carefully. Stern and Kohnert sat in front of a monitor in the telemetry room of the observatory waiting for the transmissions the detector would send after the satellite was in space.

The personnel of the Poker Flat Range were responsible for the rocket, for the propulsion, for the control of the rocket's path in space—attitude control it's called. Pat Young is Operations Controller at Poker Flat. She runs things at the range, both technical and daily management. Now wearing a white quilted jacket over her turtleneck, she was seated at the console doing her duty as operations control officer, ready to flick the switches to fire the launch.

Stern thought back to the Kennedy Space Center launch control room. Five hundred people concentrating: engineers, scientists, mathematicians, electricians, mechanics, all of their thoughts fixed on the launch. Each person studies rows of figures on the console in front of him or her, one among rows and rows of computers reporting on everything from door latches to hydraulics.

It is the same in the world of the launch at Kennedy as at Poker Flat. Will it fly? Is there some little bolt, some little connection broken loose?

Kohnert had done a launch only a few times before. It was a change from White Sands to Poker, but for him, too, it was the same world. The tension is there whether you've done 70 rockets, or one, or you're at Cocoa Beach, Florida, your binoculars fixed on pad 39B where *Discovery* is T minus 30 minutes, and counting.

T minus 30 minutes at Poker Flat. Electric life passes through the umbilical cord which connects the rocket to the blockhouse. T minus 5 and nothing can stop the rocket now. Kohnert, Stern, and all the others feel the momentum. The whole telemetry room clears. Everybody runs out into the snow.

The Taurus explodes and kicks off the pad. The rocket flames and tears its styrofoam doghouse to rags with its fins. Styrofoam in rips and hunks showers into the tundra and pine. A great fiery cloud rolls along the ground on both sides of the blast and billows up to the treetops.

In less than a second the rocket is 1000 feet up, level with the top of the observatory. By the time the onlookers can crane their necks, they start to feel the sound, a high roar rolling across the valley but louder than any thunder they have ever heard.

The rocket is screaming up, an orange streak of flame, penetrating some whispy clouds like a javelin. The flame is out. The Taurus booster separates. The rocket with its payload coasts at 20,000 feet. Above that level the second stage, the Orion, lights. It's a dot, a pin of light vanishing into the shimmering dark blue sky. It'll burn out in another 15 seconds.

Shooting through the sky, the rocket is spinning like a bullet. But the spinning has to be stopped so the instruments can focus. To solve that problem both Peggy Fleming, the Olympic Gold

Medal skater, and the rocket designers use the same technique. When Peggy Fleming wants to stop her spin, she stretches out her arms as far as she can.

The rocket will do the same thing. Inside the rocket are cables weighted at the end. One of the timers releases the cables, and they stretch out. Although the rocket is spinning at 25 revolutions per second, it stops spinning immediately. Then the gyroscope and control jets take over and keep the rocket stable.

Stern and Kohnert rush back into the telemetry room and grab headsets. The excitement has now moved from the blockhouse to the telemetry room. Stern is calling the features of the ultraviolet spectra that he sees transmitted to the computer screen, and Kohnert is retransmitting the data to Barth.

The rocket soars to 117 miles above the Earth. The planet's gravitational field has finally dragged all the rocket energy out of it. It comes to a dead stop and starts to fall. No more transmitting. Timers turn the instruments off.

Stern and Kohnert are suffused with a great feeling of accomplishment. It is the first time Stern has really been partially responsible for putting something up into the sky. UVA 84 has worked. It was up there sending down live data, doing its job. Stern felt as if he were right up there with it.

When the falling rocket was three miles above the Earth, a door flew off and an aerial unfolded and started beeping so that the rocket's payload could be tracked to where it would eventually fall to Earth and burrow into the snow.

With the mission over, Wilshusen and Kohnert packed and headed for Boulder. Stern was left behind to take care of the payload. The next day an army helicopter from Eilason Air Force Base followed the beeps and returned with the payload swinging beneath it on a cable.

Stern found the drum jammed with snow and dug out what he could, then melted the rest with a large blow drier.

Now Stern had to put into use the hands-on fundamentals he had learned from Wilshusen. He carefully removed the in-

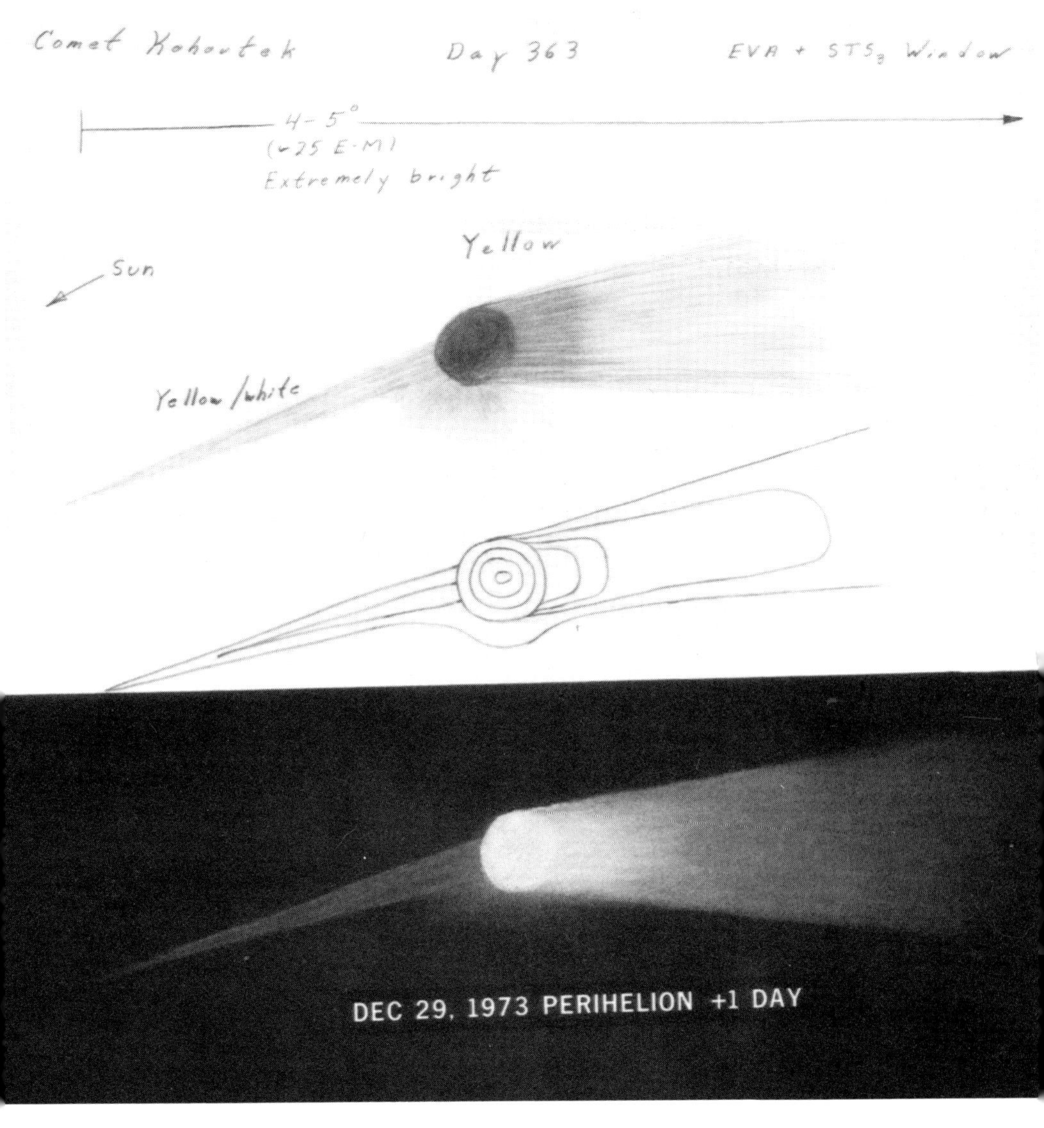

Sketches of Comet Kohoutek made by Dr. Edward Gibson aboard Skylab III, December 29, 1973. *(Courtesy of the National Aeronautics and Space Administration)*

The 200-inch Hale telescope on Mount Palomar. (*Courtesy of the California Institute of Technology*)

The Very Large Array (VLA) Radio Telescope, located 52 miles west of Socorro, New Mexico, consisting of 27 identical 82-foot reflector antennas. (*Courtesy of the National Radio Astronomy Observatory/AUI*)

Gary Emerson (*left*) and associate Rick Keen setting up special photographic telescopes. (*Courtesy of Gary Emerson and the E. E. Barnard Observatory*)

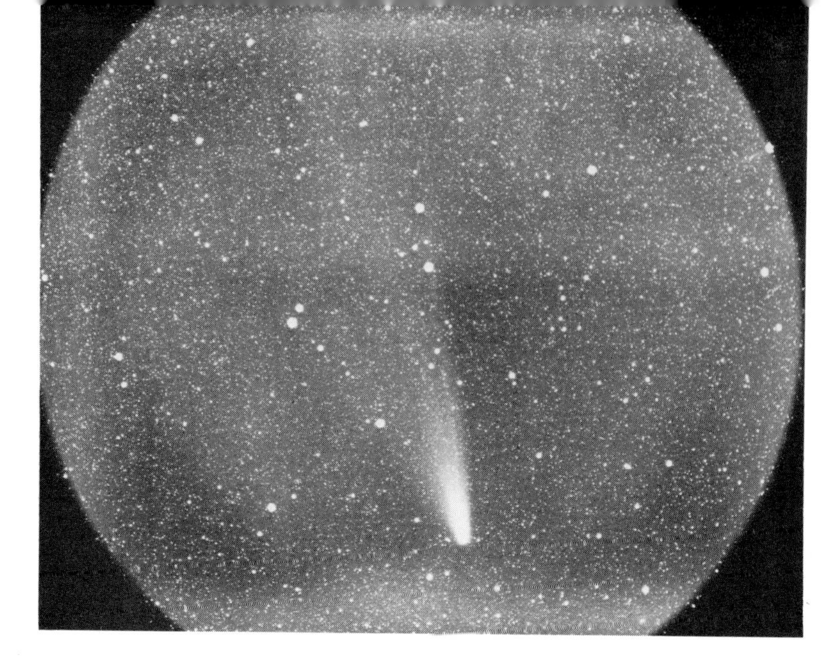

Comet West through Gary Emerson's telescope lens, 1976. (*Courtesy of Gary Emerson and the E. E. Barnard Observatory*)

A 25-minute exposure of Comet Halley taken outside Terlingua, Texas, on March 16, 1986, by Gary Emerson. (*Courtesy of Gary Emerson and the E. E. Barnard Observatory*)

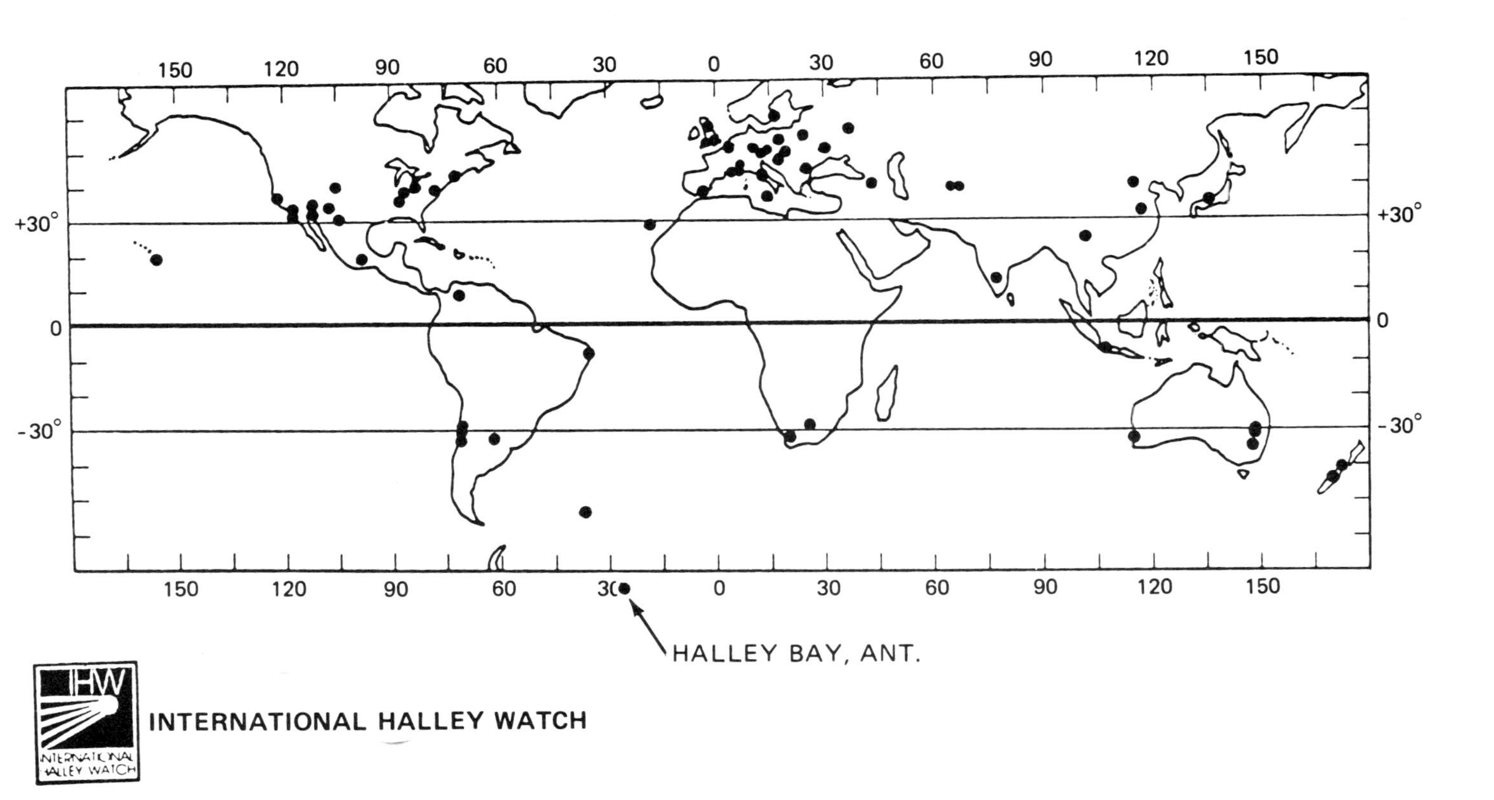

Observatories cooperating in the International Halley Watch, 1985–1986.

The spacecraft ISEE-3, later designated ICE, shown during tests at Goddard in 1977. It was launched August 12, 1978, to intercept Comet Giacobini-Zinner. (*Courtesy of Dr. Robert Farquhar and the Goddard Space Flight Center*)

Artist's conception of ICE's interception of Comet Giacobini-Zinner on September 11, 1985. ICE was the first spacecraft ever to intercept a comet. *(Courtesy of Dr. Robert Farquhar and the Goddard Space Flight Center)*

The Japanese Halley cometary probe Suisei. (*Courtesy of The Institute of Space and Astronautical Science, Tokyo*)

Suisei being attached to its launching rocket at the Kagoshima Space Center in southern Japan. (*Courtesy of The Institute of Space and Astronautical Science, Tokyo*)

The spacecraft probe Giotto assembled and in the clean-room of the European Space Agency in Cayenne, Guiana. (*Courtesy of the European Space Agency*)

The Halley multicolor camera, ready for integration with Giotto. (*Courtesy of Harold Reitsema and Alan Delamere, Ball Aerospace Corp*)

The Ariane rocket carrying Giotto on the first leg of its journey toward Comet Halley, launched at Cayenne, Guiana. (*Courtesy of the European Space Agency*)

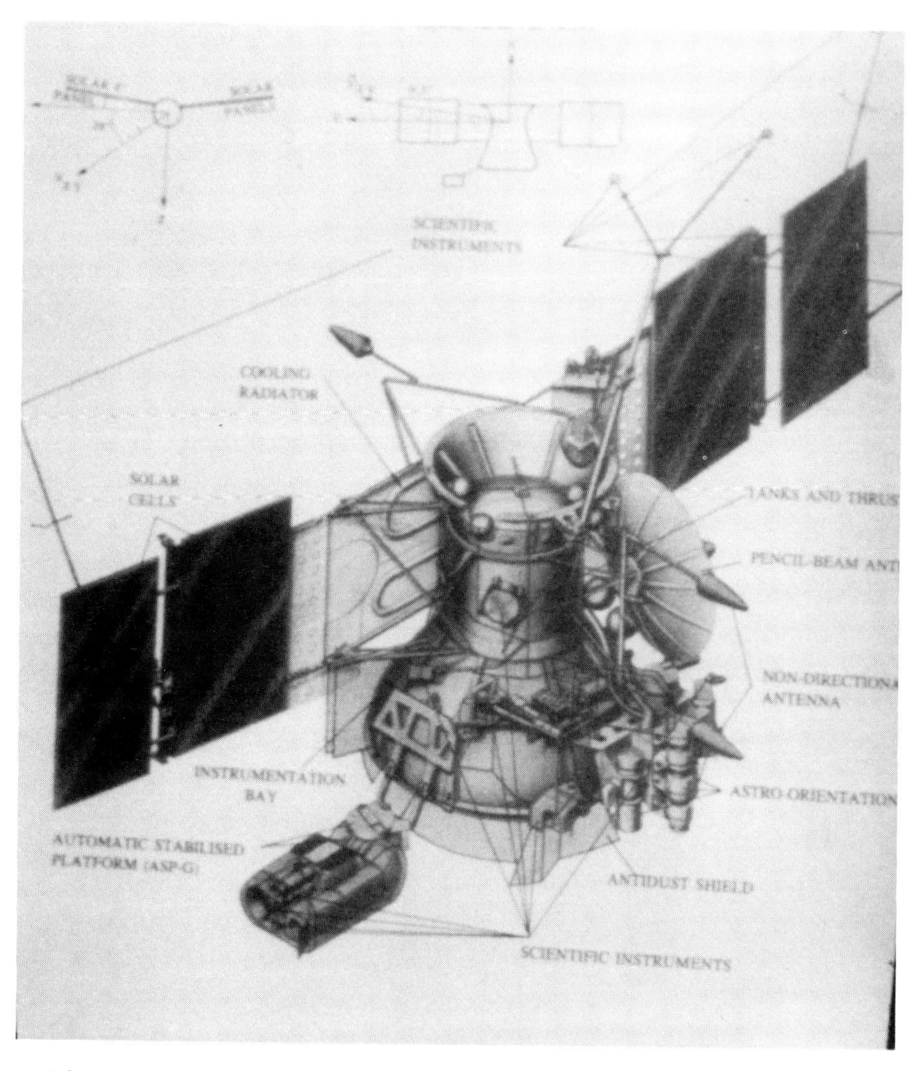

Diagram of the Russian spacecraft Vega, also sent to intercept Comet Halley. (*Courtesy of the European Space Agency*)

Spartan-Halley spacecraft in carrier, ready for testing, at Cape Canaveral. *Left to right:* Rick Konnert and Fred Wilshusen. (*Courtesy of the Laboratory for Atmospheric and Space Physics, the University of Colorado, Boulder*)

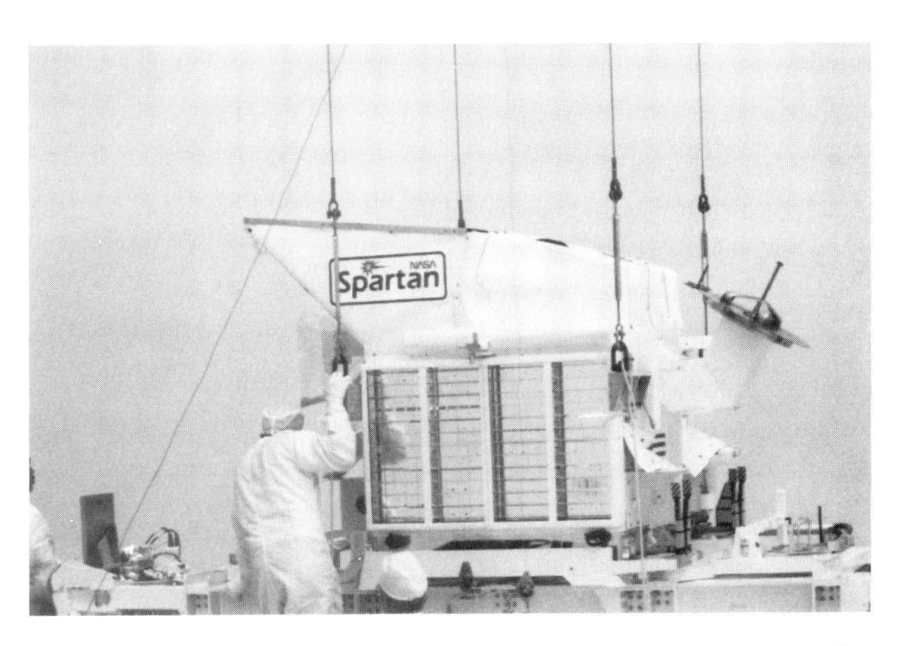

Lowering Spartan-Halley to MPESS, Cape Canaveral. (*Courtesy of the Laboratory for Atmospheric and Space Physics, the University of Colorado, Boulder*)

Cleaning the dust cover of Spartan-Halley. *Left to right:* Fred Wilshusen, Dr. Charles Barth, and Rick Kohnert. (*Courtesy of the Laboratory for Atmospheric and Space Physics, the University of Colorado, Boulder*)

Left to right: Fred Wilshusen, field team leader; Rick Kohnert, field engineer; Alan Stern, project scientist; Ellison Onizuka, astronaut; Charles Barth, principal investigator. Between Kohnert and Onizuka is the grappling device on Spartan-Halley for Judith Resnik to use after the robot arm of the shuttle deploys the satellite into space. (*Courtesy of the Laboratory for Atmospheric and Space Physics, the University of Colorado, Boulder*)

The crew of the shuttle Challenger line up for a publicity shot, with Spartan-Halley in the background. *Left to right:* Commander Richard Scobee, Michael Smith, Ronald McNair, Judith Resnik, Ellison Onizuka. *(Courtesy of Morgan Windsor and the Goddard Space Flight Center)*

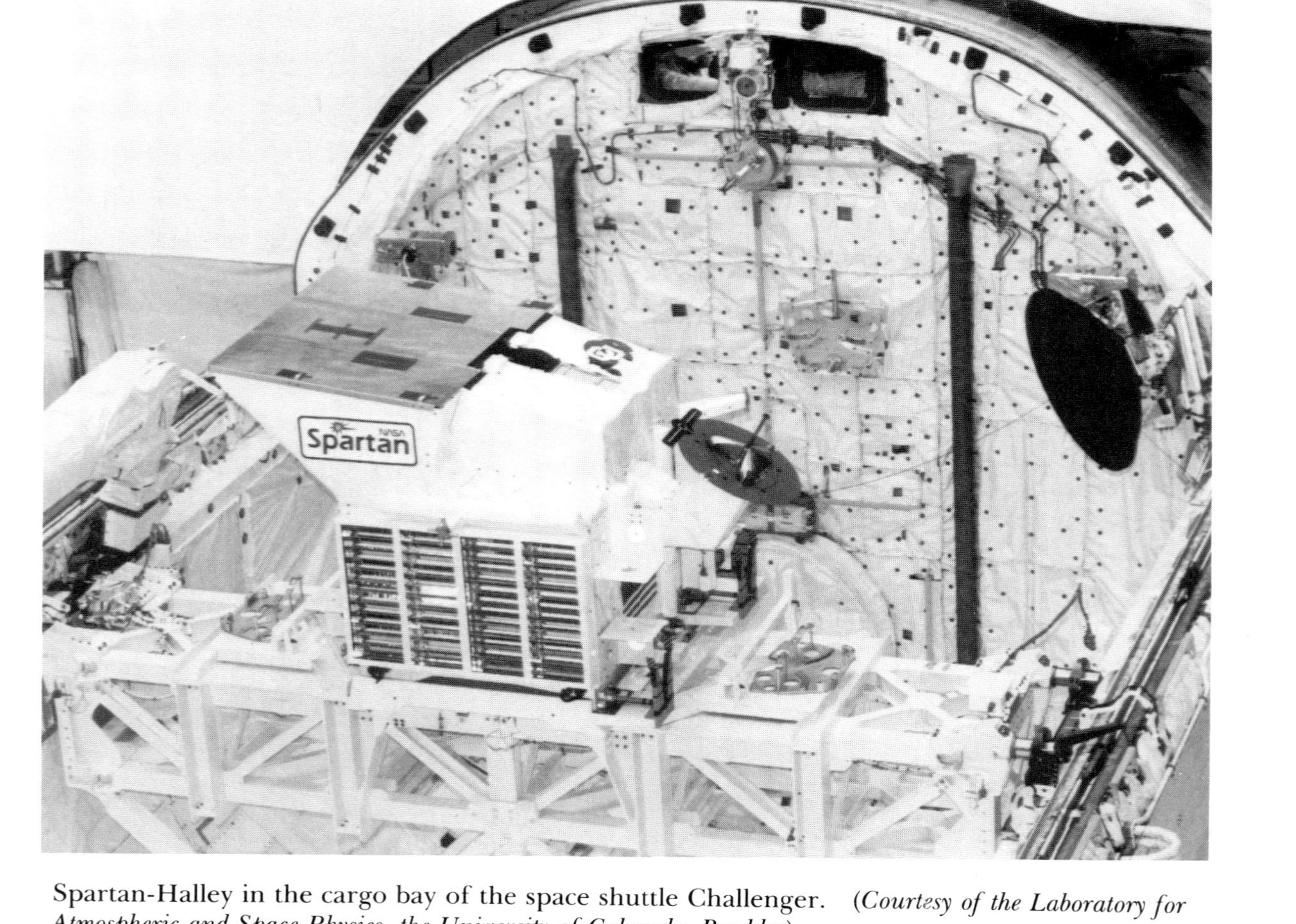

Spartan-Halley in the cargo bay of the space shuttle Challenger. (*Courtesy of the Laboratory for Atmospheric and Space Physics, the University of Colorado, Boulder*)

struments, checked their operation, then put them back into the satellite, and prepared everything for shipment to Boulder.

When Stern turned in their rented car at Fairbanks, he found they had run up a bill of $813.72 driving between Poker and Fairbanks. It was LASP's bill, but the teenage clerk had never heard of LASP. She accepted Stern's own credit card, however.

"Be back?" she asked. Stern smiled. "I'll be back." He saw the words before he said them. When he was leaving the Johnson Space Center he scratched "I'll be back" with a penknife on the under side of the Saturn V that's stretched out on the lawn as you're coming into the space center.

It had been his last night as an intern. Stern was with his girlfriend, Carole Jones. He walked up and down the path in front of the rocket nervously and explained how the engines on each stage worked, the purpose of a valve, the purpose of a piece of tubing, the physics of flight.

"What are you getting on with?" Carole asked him. Standing by the Saturn he proposed to her and she laughed and said yes.

Stern called Barth to let him know the satellite was in good condition and on its way.

"I'll be back in the office tomorrow morning," Stern said.

"Why don't you wait until Monday," Barth said. "Have a couple of days off." Stern was surprised. Barth didn't give days off, and on such a short timetable. Their bird would have to fly in a couple of years. ESA, the Russians, and the Japanese had already been at it three years. The Russian launch was planned for December 1984, ESA and the Japanese for mid 1985. Stern decided he'd spend the weekend as Spartan-Halley project scientist and continue working up his ideas for the Spartan mission.

9

The Brains of the Robot

Morgan Windsor's white hair contrasts with his black eyebrows and outdoor tan. He jogs about 2 miles every day and keeps up with flying from his World War II Navy days in his private plane.

After 25 years at Goddard he is a production manager for NASA and was assigned as mission chief for Spartan-Halley. His team at Goddard had the responsibility for the Spartan carrier, the pickup truck in space; Windsor called it a bus. The Goddard team was responsible for the Spartan's battery power, navigational equipment, and tape recorder. LASPs responsibility was care of the scientific instruments: two ultraviolet spectrometers, their ion pumps, two cameras, the power supply, and associated electronics. The cameras would take pictures of the comet's activity and the surrounding star field. Stern was excited about their possibilities.

"Our pictures'll scoop the whole bloody world," he told Wilshusen. "The Russians, the Germans, the Japanese." No question but Spartan would have the first photos of Halley from space. "They'll be in all the magazines," Stern added, "*Sky and Telescope, National Geographic, Life*."

But the instruments and carrier had to be integrated in order to operate together, LASP electronics synthesizing with Goddard. The engineers used a good metaphor, handshaking between the two teams. When the instruments would be ready for handshaking, Windsor would be in charge of the integration.

Like Wilshusen, Windsor learned the thousand ingredients and the ten thousand refinements of rocketry by the sweaty application of wrenches, soldering irons, and testing boards. Wilshusen had 10 years more experience than Windsor, but someone at NASA headquarters said the two of them together could probably wire up the Washington Monument and shoot it into orbit.

Neither man was a theoretical scientist, but by drawing on their practical experience they were able to design all the circuitry for launch, attitude control, instruments, everything on Spartan-Halley. When a problem needed aeronautical theory, Windsor deferred to others and let them solve the problem without interference.

During an early conference at Goddard, Windsor introduced John Laudadio. Laudadio's job was to make sure that everything about Spartan inside and out conformed to NASA safety regulations. A lanky, mechanical engineer who had been with NASA for 20 years, Laudadio was polite and insistent at the same time, so no one mistook his politeness for weakness. Knowing the safety reports engineers and technicians continually filled out were resented, Laudadio helped everyone with them with a quiet reminder of their importance.

"Our number one responsibility is protecting those astronauts," he said. "We can't accomplish the mission if we can't do that." He handed Sam Jones, LASP chief engineer, a book 3 inches thick that regulated connections, bonding, voltages, lubrication, paint, sharp edges—everything that had to be controlled to prevent accidents.

Comet Halley had given the mission its timetable, and Windsor told the LASP people they would have their instruments ready to integrate with the carrier and at the Goddard Space Flight Center by January 1985.

Nobody at LASP admitted to panic, but the schedule would definitely be a challenge to their best efforts. It could take a year just to design the instruments. Then the instrument makers had to construct the instruments, the electric assemblers had to put together circuit boards. A single circuit board might take a month of 8-hour days. Then the instruments had to be positioned for the carrier.

Later there would be 10-hour days, then 12-hour days; then 12-hour days for weeks at a stretch and the fatigue that sort of pace engenders are not compatible with measuring connections to the 10 thousandths of an inch so that power flowed through thousands of wires, from batteries to instruments, without shorting out or sparking an explosion.

Basically, the reusable Spartan carrier was an aluminum box 4 feet wide, 5 feet long, and 3 feet high, not counting projections like the gas jets or the grapple for the Shuttle's cargo arm. Spartan was as self-contained as Skylab, but as a space bus it was too small even for a crew of hobbits. Unlike Skylab, once deployed in its orbit, Spartan would be out of contact with anyone, hobbit or human.

This self-contained minibus would begin its voyage loaded on the shuttle thundering off from a Kennedy launch pad. Two days and nine hours later, as the shuttle orbits the Earth about 200 miles up in space, Spartan would still be in the cargo bay, as one of the astronauts, the Mission Specialist, turns on Spartan's power. When satisfied that everything's "go," another astronaut picks up Spartan with the robot arm (officially the remote manipulator system arm), which hands it out into space and drops it off.

Within 30 seconds Spartan turns on its side, performing a "pirouette," as an imaginative planner called the maneuver. Now it's in the correct position to be shaded from the Sun, and it starts orbiting the Earth every hour and a half, moving a little faster than 17,100 miles per hour, about the same speed at which Skylab circled the Earth. This is a standard Earth orbit for spacecraft, not fast enough to shoot into space, not slow enough to fall back into the atmosphere.

After Spartan locates itself in space then for the next 44 hours, orbiting the Earth 28 times, it scans Comet Halley, which is about 220 million miles across space. It takes pictures of Halley, rests during its night on the other side of the Earth, emerges into day, measures, and photographs again. Every fifth orbit, Spartan checks on itself to make sure it's not out of position.

After four days, when Spartan's mission is completed, the shuttle orbiter returns and the robot arm grabs it. Once Spartan is safely latched in the cargo bay, the mission specialist checks the records of the mission on Spartan's tapes to make sure the spacecraft operated as programmed out there all by itself. After receiving the OK, the mission specialist turns Spartan off and the shuttle goes on to other missions. Two days later, the shuttle returns with Spartan to Earth.

Stern spent that weekend Barth gave him after the UVA 84 launch in Alaska working out details that would prevent Spartan from failing to accomplish its mission. Then he returned to the LASP engineering building, known simply as "55th Street" because that's where it's located in Boulder.

He joined Wilshusen, Kohnert, Bob Leyner, who directed the mechanical operation of the instruments, and others in Sam Jones' office. None of them could miss the huge schedule 5 feet long and 3 feet high that Jones had taped on the wall. There were vertical lines indicating dates, horizontal lines indicating the start and completion of projects. There were arrows pointing to special dates, reviews of progress, tests of equipment. There were lines for the construction of the spectrometers and components, the microprocessors. Jones wanted the lines and arrows to establish goals, not to intimidate. Still the shortness of time *was* intimidating.

Jones' office was on the first floor of 55th Street, the only outer-inner office complex. His outer office had a worktable, and the opposite wall was a blackboard. The inner office was more executive-like, with desk, chairs, carpet, computer: it was used for special one-on-one conferences.

The entire cement brick building had once been a warehouse. It was now painted brown, and from the southwest angle was surrounded by trees and lawn, giving the impression of a park. The east side of 55th Street still looked like a warehouse: cement blocks still gray, an asphalt parking lot, a loading dock, and a sliding black steel door where trucks used to bring their cargoes.

The mills, drills, lathes, and other machines with their control gauges were on the warehouse floor. Here the instrument makers drilled and cut aluminum, which is easy to work compared with titanium or high-temperature alloys like tungsten. Most of the offices surrounded the machine floor. They were walled out of plywood and coated with a transparent protective coat, so everything looked temporary and new after 17 years. Most of the offices were tiny. Even Wilshusen's, for example, was long and very narrow, with not much room between his drafting board, his stool, and the wall behind his back.

The scientists and engineers of LASP crowded around the table in Jones' outer office redesigning the payload for the Spartan carrier. It was like a family working out the plans for its new house. There was only so much space to play with and a lot of stuff to fit into it.

The two LASP spectrometers were separately boxed units with the boxes placed so the spectrometers and the cameras looked out at the comet together.

Nothing in the payload could be very far from anything else. Where to locate the batteries that supplied the power? And the wiring from them to the instruments? The networks of wires and harnesses (braided bundles of wires) couldn't interfere with each other, and all of them had to be interfaced with the wiring Goddard laid out for navigation. Move a navigational instrument, the star tracker, for example, and you had to move something else, then reroute the wiring to accommodate the moves.

Leyner had one of the machine people construct wooden models of the instruments, which he moved around like chess pieces. When Leyner moved a piece a little to the left, Wilshusen showed what that change would do to the route the wiring had to follow. In all his decisions Wilshusen was guided by the distillation of all his past satellite experience. On any ticklish problem, the people at both LASP and Goddard asked his advice and made a decision on the spot.

Decisions on Spartan-Halley were not made by high-echelon management in a 30th-floor penthouse. They were made by the scientists and engineers working on the job. Spartan was a family satellite. LASP was a family, and everybody who worked on Spartan felt personally involved with it: Barth, Jones, Wilshusen, Stern, the instrument makers, the women who soldered the circuit boards.

Spartan was the result of teamwork. Sometimes a scientist worked on optics and a mechanical engineer integrated the electronics, bringing the mechanical and electrical together. The skills of scientists and engineers fused into Spartan-Halley, the little spacecraft that could.

Big outfits like Martin Marietta, Rockwell, or General Electric have 100 to 200 times the people on their projects than were available to work on the Spartan-Halley mission. At most there were 25 in Boulder, 35 at Goddard.

There was a special personnel syndrome in Boulder, which is located at the base of the Rockies, namely the outdoor syndrome. Many of the LASP engineers have an out-in-the-country address, a star route number, a small house on Peak Road, or a rustic cabin on Winding Canyon Trail.

In addition, the engineers' sports tend to be outdoor activities, rock climbing, skiing, camping. Paul Bay, one of the instrument makers, is a water rafter and kayacker. Gravity explodes in the spasm of water. Like space, it's the escape from Earth into an alien environment.

Mountain climbing, a different kind of escape, was not restricted to Boulder but also extended to NASA administrators. One of the most experienced mountaineers in the United States was Tim Mutch, director of space science at NASA in 1980. He led a Himalayan expedition to Mt. Nun, 23,410 feet, the highest peak on the eastern Kashmir massive, about 350 miles from New Delhi, India. Mutch died in an attempted ascent.

Sam Jones, a mountain photographer and the LASP supervising engineer, believed the engineers and instrument makers didn't need motivation indoors at 55th Street any more than they needed it outdoors in the mountains.

"You can't lie to a spacecraft," he said. "Space will find you out." Jones meant that work on a spacecraft has to be exact. Yvonne West and Beth True assembled the electronic circuitry for most of the instruments of Spartan-Halley in the 55th Street clean-room. The wiring connections were so minute they had to be soldered under a microscope. West and True were able to work only a couple of hours at a time. One had to spell the other so the first could rest her eyes and fingers. True was asked if she repaired her own electrical appliances in her spare time.

"I looked in the back of my TV once," she answered. "I was disgusted. It was a bunch of crap." Her comment expressed the difference between the exactness demanded by space engineering and the manufacture of household appliances, the difference between a work of art and a "bunch of crap."

Jones knew he didn't have to check up on anything West and True assembled, and he applied the same hands-off attitude to those working on the Spartan instrument computers. Computers demand perfection. Neil White constructed the computer that operated the Spartan observatory. He was not one of the outdoor people. His hobby was computers. His idea of recreation was designing and constructing supercomputers. At his home there is always one in the process of evolution.

The Spartan computers were actually microprocessors, the guts of a computer, the chips and circuits without keyboard, screen, or printer. White followed the usual spacecraft practice and assembled two models of his microprocessors, a flight model, the one that would fly on the spacecraft, and an engineering model, identical to it but not certified for flying, which he could use for testing programs and electronics. In designing the microprocessor he doubled its memory chips, placing one set on the right and one on the left, keeping to the redundancy principle to ensure that Spartan-Halley wouldn't be left floating in space with amnesia.

White also had designed some very special advanced chips for his microprocessors. Being advanced, the design would be expensive. The LASP budget was low and getting lower, so Jones told White to give National Semiconductor a call to get some sort of estimate.

White called and told an NSC executive just the kind of chips he had to have.

"Great," the executive said. "We have them, but you can't get them for six months."

"Six months!" said White. "We're going to launch in a year. We need them now."

"Tell you what I can do for you," said the executive. "I can get what you want off the production line, but it'll cost you."

"Money? We work for NASA. We haven't got any money."

"NASA? What are the chips going to be used for?" the executive asked.

"It's a satellite going to Comet Halley on a shuttle."

"Comet Halley! The shuttle! Buddy, you sign a release that we can advertise that our chips are going to Comet Halley on the space shuttle and you can have them in two weeks and they'll be free." A jubilant White paraded into Jones' office with a quite unexpected victory for NASA, Halley, and LASP.

Greg Palmer, a recent University of Colorado graduate, waiting to be called into the Navy, was the programmer for White's microprocessor. He already had a reputation as one of the best University of Colorado hackers. He wanted to be sure what his program had to accomplish and built a little wooden model of Spartan with two dowels extending out the sides. He and White crowded into Palmer's cramped office and rolled the model left and right, up and down, as the craft would move in space.

"Over where you are is the Sun," said Palmer. "Spartan rolls this way." Then he rolled the model in the "pirouette." Directions for all the maneuvers had to be in the program. White then formed an angle with his arms.

"Over here is Halley." The program had to compensate for possible errors. Stern had told them how many degrees the spectrometers had to scan to measure the coma of the comet. The program called for scanning a little in front and a little in back just in case.

Palmer understood the problem and withdrew. Like many programmers, he became a recluse when he built up a program.

"Don't bother me," he said. "My program'll work."

Another microprocessor, the navigational microprocessor, directed the Spartan where to fly in space. This one was Goddard's responsibility and had been programmed by Jim Watzin, their flight systems engineer.

When the shuttle dropped off Spartan, the little spacecraft's sunsensors would start searching for the brightest object in the sky, that brightest star, the Sun, and Spartan would lock in on that. But the spacecraft still needed a second point, another star. But which one?

Many astronomers like to use Canopus in the southern constellation Carina. Not only modern astronomers, but early explorers Columbus, Vasco da Gama, and Magellan were partial to Canopus as well. The third brightest star in the sky after the Sun, Canopus is always very close to a right angle from a line established on the Sun.

So after Spartan was locked onto the Sun, its star tracker, a central instrument in its navigational equipment, would locate Canopus. Fixed on the Sun and Canopus, Spartan would be ready to find where Comet Halley is at any precise moment.

No one knew which precise moment that would be until the shuttle was launched, so Palmer had to program more than 1000 positions where the Comet might be during the two months of the mission window, January 8 to March 7, 1986.

When the time came, the LASP computer would tell the navigating computer where the comet was located. The two computers had to work together like Scotty and Captain Kirk on *Star Trek's Enterprise*. Scotty, the Goddard computer, would ask, "Where are we now?" Captain Kirk, the LASP computer, would tell him.

"Where are we going?" Computer Scotty would ask. Computer Kirk would tell him. To get there Scotty would give the word to jet out just the right amount of argon gas stored in two tanks and under pressure.

At one of the many meetings that Windsor called between LASP and Goddard, Stern and Watzin discussed another navigational problem. Until Spartan, which had been taken off the *Discovery* flight, had been reassigned to a another shuttle, no one would know for certain the exact altitude of Spartan-Halley when it would be deployed. The Spartan orbit in space lasted about 90 minutes. What if the shuttle was higher and the orbit took a bit longer? Let's say a bit is 10 seconds. In 26 orbits Spartan-Halley would be off by 4 minutes, and in 4 minutes it would be more than 1000 miles off and everything would be in the wrong spot, the Sun, Canopus, and Comet Halley.

Windsor challenged Stern, in his job as project scientist, to come up with a way that all the pertinent information could be entered into the Spartan-Halley computer after the carrier had been locked into the shuttle's cargo bay and the shuttle was in orbit above the Earth. Stern became as much a recluse in his office at Boulder as Palmer and solved the problem.

Stern figured the mission specialist for Spartan-Halley would need to enter five pieces of information if Spartan was to locate Comet Halley: the month, day, time of day, time expired since launch, and the exact altitude.

Mission specialists communicate with payloads in the cargo bay with something that looks like a large pocket calculator, a "get away special" Autonomous Payload Controller (APC) used to switch on the power for the experiments, turn it off after the mission is completed.

As soon as the shuttle was launched and had reached its orbit, mission control at Johnson (and the astronauts, too) would know the date, the time, and the exact altitude. Then and only then could those five bits of crucial information be entered into the navigational microprocessor so it can position Spartan accurately. And only the mission specialist could enter that information, because Spartan would be out of communication with the ground.

Computers translate the signal from the key the operator punches into a binary code, a series of switches that are either on or off. For example, if you want "5" on your computer screen, you hit 5 on your keyboard; in a millisecond, chips inside the disk drive close one switch, open a second, close a third, and 5 appears on the monitor screen. The number 6 on your keyboard activates a different series of switches in your disk-drive.

The mission specialist's APC does not translate his signals automatically into binary numbers. He has to go directly to the switches. He has to do his own translating into binary as if he had to use a new language. A mission specialist on a shuttle flight has enough to do preparing for a mission without learning binary.

Someone in mission control on the ground could figure the binary ahead of time—Stern thought it would probably be himself—and teleprint it up to the shuttle in a series of H's and L's for each command. H to send high voltage through a relay (no switches available in the microprocessor), L to send low.

Reading the rows of H's and L's from the teleprinter, the

mission specialist would enter the five pieces of information that Spartan needed. A certain series of H's and L's would mean "The date is January 17." A second group of H's and L's would mean, "The time is 1032."

The APC is supposed to operate simply, but Stern knew his system for Spartan-Halley would really tax the system, and to its limits. Stern, therefore, built some checks into the system. The microprocessor would signal the H's and L's it received back to the mission specialist to make sure the series had been transmitted correctly. Other checks were built in as well.

Stern was excited about this system. It would be time-consuming for the specialist and would crowd his already tight work schedule during the flight, but Stern knew his system could accomplish its task. When a shuttle crew was finally assigned to Spartan-Halley, however, Stern would have to sell it to the mission commander and the specialist. He hoped he could do that.

10

One Bird Flying

Wilshusen, Jones, and White did not have to be reminded that as they were designing the layout of the Spartan instruments and constructing the microprocessors, the European Space Agency, the Russians, and the Japanese were already two years ahead of them. ESA had started its planning and design phase even before Danielson had recovered Halley on the Big Eye on Mt. Palomar, October 16, 1982.

In Bristol, England, British Aerospace had already constructed three models of the Giotto spacecraft. One was structural. They cooked it, froze it, and shook it up to make sure the model would survive the rigors of space and the shock of launch. Another model was electrical, the engineers making sure all the circuits were compatible, that one didn't reject the other, short it or somehow arc. The third was "it," the flight model, the real Giotto that would go to Halley. Into it the engineers incorporated the toughness of the structural model and the compatibility of the electrical system.

The spacecraft, the barrel adapted from Geos, was a pile, 5 feet high, of four flat metal bagels that were actually platforms: one platform was for instruments; two were for the attitude control system; and the fourth was a radiator. The hole in the bagels was filled with the Earth boost motor, other motors of various sorts, the mechanism to control the dish antenna, and other structures.

A variety of sensors located at different places on the spacecraft gathered the data and sent it to the instruments. The 10 experiments were fixed in a circle ready to receive the data, analyze it, and measure it. Giotto had no way to store information, so all the data collected by its instruments and camera had to be transmitted down to the European Space Operation Center at Darmstadt, West Germany.

Roger Bonnet, ESA's director of science, had announced that Giotto would fly into the coma of Halley as closely as possible to the nucleus. When the American newspapers reported on the progress of Giotto, they always headlined the story: "Germans build kamikaze spacecraft." However, Rudeger Reinhard, the project scientist for the Giotto Mission, wasn't about to have the spacecraft clawed into pieces or the scientific instruments knocked out if he could prevent it.

A scientific working group decided that shielding the spacecraft would best protect it from the ions and dust particles. And who could best design the shield but Professor Fred Whipple of the Harvard-Smithsonian Center.

Whipple designed a double shield system. The front shield was aluminum, about 4 hundredths of an inch thick and covered with a coating of gold about 100,000ths of an inch thick. Gold protects against the penetration of ions three or four times more than aluminum. Dust particles, however, might penetrate the gold and even the aluminum beneath, but the particles would be vaporized in the process.

Then the vaporized dust particles would find themselves in a gap of about 10 inches. The vapor would expand, lose energy, and not be able to progress any further. It would be then up against the inner shield, almost a half inch of Kevlar, the stuff of bulletproof vests.

The two shields weighed only a little more than 100 pounds. Judging from his knowledge of the composition of comets, Whipple thought the spacecraft had a good chance of surviving a flight within a few hundred miles of the nucleus.

Giotto wasn't planned to be a kamikaze spacecraft, nor was it built only by Germans. Reitsema and Delamere at Ball Brothers, working on the camera, were only two of about 37 Americans who worked on Giotto as co-investigators for the experiments, and there were more Americans represented than any other single nationality. The other nationalities involved were Belgian, French, West German, Dutch, English, Italian, Irish, Australian. Giotto was as truly international as art, as Giotto di Bondone, as Van Gogh, as Jackson Pollack.

Uwe Keller, the principal investigator for the camera on Giotto, as he had been for the canceled Halley-Tempel 2 probe, was impressed with the improved design Reitsema and Delamere submitted. They had compensated for the rotation of the spacecraft during its flight and the speed with which Giotto would be flying past the nucleus. The camera used neither film nor shutter but two CCDs (charge couple devices) which recorded the light images and stored them for transmission to mission control. The operation was controlled by three microprocessors: one to aim the camera, the second to work it, and the third to transmit the recorded signals.

If Giotto flew even within 800 miles of the comet nucleus—and it was planned to fly it within 400 miles—the camera would show features measuring less than 100 feet. However, the telescope of the camera could not be protected by the shields. The length of the tube, about 10 inches, would provide some protection from the dust particles, but the length wouldn't be enough by itself. In addition, at the bottom of the tube was a mirror of special metal set at a 45-degree angle, which reflected the light to the camera without the dust particles bouncing around the corner.

The camera was also international. Keller had traveled all over Europe raising funds for it. He got some money from the German government, some from the Max Planck Society, and some from NASA. The telescope came from France and was financed from Paris. The mirror and baffling within the telescope was Italian. The two CCDs were manufactured at Texas Instuments and tested in Delamere's office in Boulder.

All the experiments were as international as the camera. Continental European, British, and Irish laboratories and institutes worked closely with American counterparts. Scientists from Heidelberg, Germany; the University of Bern, Switzerland; the University of Kent, England; Toulouse, France; Bari, Italy worked with scientists at the University of Texas, Dallas; the Jet Propulsion Laboratory; Lockheed, Palo Alto, California; the Space Science Laboratory, Berkeley, California; the University of Florida, Gainesville; and many other institutions.

The dust particles shooting from the nucleus of Comet Halley would be counted, measured, and analyzed by two major systems: first, a particle impact analyzer to report their chemical composition; second, a dust impact detection system (DIDSY) was actually six sensors under one microprocessor, five sensors on the front shield and a sixth on the rear just in case particles managed to penetrate.

Spectrometers would analyze the composition of the coma, and magnetometers would measure the magnetic conditions where the solar wind impacts on the front of the comet, forming a bow wave.

For Giotto to rendezvous with Halley March 13, 1986, its launch window had to be the first two weeks in July 1985. British Aerospace had to deliver the flight model to Toulouse, France, by February 1985, and it did.

Some parts of the camera dawdled into the Max Planck Institute at Lindau at a frustrating crawl. Keller and his staff were supposed to integrate the parts, then integrate a functioning camera into the spacecraft at Toulouse.

The Lindau engineers finally received all the parts and assembled them. Unfortunately the assembled camera wasn't getting correct instructions from the computers, a problem with software. Keller called in Delamere from Colorado as a consultant on programming. After all, the camera was operated by three microprocessors, and Delamere had more experience programming microprocessors than the rest of the camera staff.

The frantic Halley scramble common to Spartan now began at Lindau. Delamere, Keller, and engineers worked around the clock, seven days a week. Giotto, with other instruments but without the camera, was shipped from Toulouse to Kourou, French Guiana, the Ariane space launching site.

The camera, when Delamere and Keller got it functioning, went from the clean-room at Lindau to the clean-room at Liege, Belgium, for testing in a vacuum tank. Then the camera was flown, its purity preserved in a plastic womb, to the clean-room of the payload assembly building at Kourou.

Delamere and the others from Ball Brothers returned to the mountains of Colorado. Keller and others from Lindau went to the steamy jungles of the South American coast.

French Guiana is known as the location of Devil's Island, but the penal colony, finally closed down in 1947, is 15 miles off shore from the small capital of Cayenne. French Guiana was chosen as a launching site because it is only 5 degrees from the equator. Because the diameter of the Earth is largest at the Equator, the rockets get a free lift of 1650 feet per second, 6 percent of their orbital velocity. The Centre Spatial Guyanais, the space center, was constructed at Kourou, 25 miles to the north of the capital. The TV cameras that follow the launch were mounted only a few yards from a crumbling wall of Royale, part of the old French prison complex.

The security of the space center was maintained by the French Foreign Legion. It was disappointing that they no longer wore the blue and red uniforms made famous in late night TV movies but rather olive green ones with shorts. They still wore the white pillbox hat, but without the neck cover that fluttered so dramatically as Gary Cooper rode out across the desert.

In the space center, the ancient rain forest had been cleared from the airfield and the launch area, but the wet, heavy air wrapped around the North European visitors like invisible steam.

In the payload assembly building everyone was suddenly back in Lindau. The air was cool and the humidity controlled as in all clean-rooms throughout the space world. In the center of the room was Giotto, vertical, the triangle sensor support above the angled dish antenna.

The instrument platform of Giotto had been opened. Keller directed the camera placement and its final testing. He discovered it still had a problem. It wasn't that it would take bad pictures, but during its flight and encounter the controllers would have to babysit for the camera more closely than originally planned. No time to remove the camera from the spacecraft, break it apart, fix some wiring, reassemble it, replace it, and reintegrate it. So, babysit.

Problems were continual, not only with the camera but also with some of the other instruments and with the telemetry. Nothing is ever perfect, but Giotto was operable and OK'd for flight.

Giotto, enclosed inside an aluminum capsule, became a projectile at the top of the rocket, a white, 211-ton tower. The tower moved on a railroad platform from the assembly building a half mile to the launch pad.

The Giotto team went to the control room, 7 miles away. It was July 2, 1985. The launch window was at its start. Isolated in the control room, the scientists, engineers, and controllers watched the three stages fire, the size and sounds of the explosion reduced on a TV screen. Everyone was buoyant and elated. Others in the bunker joined them, laughing, slapping, hugging. Giotto was on its flight around the Earth.

In the United States the Fourth of July holiday had begun,

as usual, a bit early. Reitsema was celebrating in a cabin outside the small mountain town of Dillon, Colorado. It was 9 o'clock the night of July 2. It was Reitsema's turn for the beer run. He turned on the radio in his car and heard the news, car accidents on I-25 out of Denver, some political drivel. The announcer added, "Oh, they launched the Giotto spacecraft today." Then the announcer went on to the balloting for the All-Star baseball game the next week.

The announcement of the launch brought Reitsema back to the world he was vacationing from. He was surprised. ESA got it off early. Reitsema grinned to himself. With all those problems he knew they were having at Kourou, he had thought they would take the whole window to get the launch completed.

Delamere was on a rock-climbing expedition in the mountains of southwest Colorado. When he returned home, his wife told him Keller had called from French Guiana. They had a perfect launch.

Each of the three stages of the Ariane rocket signaled its flight in detail to the computers in the control room at Kourou. Lines of figures traveled up the computer screens all during the launch, and later they were printed out to be examined and appraised.

When the engineers of Arianespace looked over the printout, they discovered a valve in the first stage had malfunctioned but late, after the second stage was already carrying Giotto into the sky. If this had happened earlier in the sequence, Giotto would have been dumped into the Atlantic Ocean. Actually on the next launch of Ariane 1, the same failure occurred, and it occurred early. That payload wound up in the sea.

The Giotto launch was a lucky success. Most of the scientists and engineers returned to Europe, but they left a small crew behind. Some mission control equipment had been shipped to Kourou to help monitor the launch, and this crew repacked it for the return to Darmstadt. Until this equipment was set up again, Giotto sailed around the Earth, isolated, a bird alone. When the equipment arrived at Darmstadt, the engineers quickly installed it, and once more communication was established with Giotto.

Giotto circled Earth three times. At exactly the right moment the flight director signaled the booster engine to fire its solid fuel. He wanted to have an additional advantage by moving the spacecraft out of Earth orbit when the direction of Giotto was exactly parallel to the Earth's motion around the Sun, thus using the Earth the way Farquhar had used the Moon, to increase spacecraft speed. Now Giotto was soaring 48,000 miles per hour toward Halley. Paralleling the Earth's orbit but inside it, the revolving spacecraft moved around the Sun, and the controllers stopped the dish antenna from spinning with a system something like the "Peggy Fleming despin" on UVA 84.

Keeping itself aimed properly, Giotto signaled mission control its position in space. The controllers corrected the flight a little. Actually, at this time they didn't know the exact position at which Halley's nucleus would be at rendezvous, and they wouldn't know it until the Vegas had sighted the nucleus during their dive into Halley's coma in March 1986.

Ground stations around the world were now getting radio signals from Giotto, including Carnarvon on the west coast of Australia, and also from the stations on NASA's deep space network: Goldstone, California; Canberra, Australia; and Madrid, Spain. The controllers at Darmstadt switched on the camera and the other instruments again and again to test them. Everything seemed to be working, if not perfectly, at least below the level of desperate pencil chewing and coffee gulping.

Reinhard was happy about the way the flight was going. In his press conference, though, he kept saying, "We are not sure Giotto will survive the encounter," and added, "We hope it survives long enough to send us good data." So the mission's kamikaze image stuck in the public mind.

11

Two Birds Flying

In Moscow the building that houses the Institute of Space Research, the IKI, is located on the southwestern edge of the city down Leninsky Prospekt, past the towering statue of Yuri Gagarin. It has the look of an older American urban university, but some academic buildings look the same wherever they are: in Wuhan, China, or Buffalo, New York, gray, cement-block walls with featureless windows.

This building houses not only the administrative offices of IKI but also conference rooms, modern laboratories of various sorts, and other rooms cluttered with electrical testing boards and computer consoles. The Institutes of Geochemistry and Analytic Chemistry occupy other sections of the building.

The office of Roald Sagdeyev, the Director of the IKI, was modern like the laboratories and conference rooms but still not typical of the skyscraper office of an American executive. It opened onto a small greenhouse, which gave it a distinctive wet-earth smell. Sagdeyev, usually volatile and open, never explained to visiting scientists why it's there.

In 1981 Sagdeyev began turning two Veneras into two Vegas, redundancy in its purest form. He faced some of the same problems that Barth and the LASP engineers had encountered constructing Spartan and that British Aerospace had constructing Giotto. Essentially all satellites have to perform the same tasks, so therefore they create the same headaches.

Sagdeyev's spacecrafts were the largest ones to go on a Halley mission. The Vegas were over four times the size of Spartan, three times the size of Giotto. They were shaped like monumental egg cups 11 feet in diameter and 16½ feet tall. They weighed 4½ tons.

The Vegas were planned for a longer flight than any of the other craft. Spartan was scheduled for 29 trips around the Earth, 200 miles above it, a trip of roughly 760,000 miles in four days. Giotto had to travel 93 million plus miles in eight months. The Vegas had a 14½ month trip to Venus and from Venus to Comet Halley, roughly 220 million miles. Tracking problems, guidance problems, and communication problems would be multiplied over those of the other spacecraft.

The Vegas would leave Earth on Proton SL-12 rockets. The Proton, 195 feet tall, is a three-stage rocket with six strap-on boosters that would develop a thrust of over 3 million pounds. The Proton is about the size of our Titan.

Sagdeyev built his Vegas on good foundations. The Veneras are excellent vehicles with 20 years of success withstanding the vacuum and temperatures of outer space, and the pressure and temperatures of Venus.

While Sagdeyev was negotiating with the Soviet Academy of Sciences and designing the Veneras for the Halley experiments, Venera 13 and 14 had both successfully landed on the 900-degree surface of Venus. Both spacecraft transmitted TV pictures in two colors, dug up soil samples, pulled them back into the spacecraft, analyzed their composition, and sent the results back to Mission Control northeast of Moscow.

But dropping into the atmosphere of Halley is different from dropping into the atmosphere of Venus. The Soviet engineers added the necessary dust shields, but if two were good enough for Giotto, four would be even better for Vega. Some particles, the designers believed, might penetrate the first shield, some even the second, but any that might get through the third would be vaporized and would not be able to penetrate the fourth and enter the interior of the spacecraft.

The television system of the Vegas was considerably more sophisticated than that used on the Veneras and as international in provenance as the Giotto camera, a combination of French optics, Hungarian electronics, and Russian hardware all mounted on a Czechoslovakian platform, which was attached to the platform of the spacecraft but suspended outside.

This television system consisted of two cameras with telescopic lenses; the first camera was to function as the Vegas' scout, able to see far ahead, able to pick up the comet in the distance and track it as the Vegas approached. The second would be able to provide a closeup shot of Halley's nucleus while the spacecraft was still 6000 miles away.

The light coming into each camera divided into two paths for video transmission and various scientific analyses. The system was equipped with four CCDs and several microprocessors that moved the platform, aimed the cameras, collected the data, evaluated it, processed it, and transmitted it to mission contol.

The Veneras all carried solar panel wings extended outward like those on Skylab, not wrapped around the body of the craft like those on Giotto and ICE. To power the 14 Halley instruments, the wings on the Vegas had to be 30 feet long, twice the size of those on the Veneras.

Sagdeyev assembled the hardware and software for these experiments from various countries outside Russia, eight in all: East Germany, West Germany, France, Hungary, Bulgaria, Holland, Austria. Among them were three kinds of spectrometers, ion analyzers, and an Austrian magnetometer called MISCHA (the Russian diminutive for "Michael" or "bear").

The eighth contributing country was the United States. With the help of a NASA grant, John Simpson, professor of physics at the Enrico Fermi Institute of the University of Chicago had invented a dust particle counter and analyzer more sensitive than any previous instrument and capable of detecting particles as small as one-tenth of a trillionth of a gram.

But where was Simpson going to use it? He had it perfected by 1983, and asked Bonnet if the ESA could use his analyzer on Giotto. The instruments for Giotto were already set, but Sagdeyev heard about the counter and analyzer and told Simpson he'd make room for two of them on the Vegas. A slight hurdle had to be jumped. The 1972 agreement for cooperation between American and Soviet scientists had lapsed in 1982. The Reagan administration was not about to renew it in spite of a Congressional resolution to do so. Thus the analyzers could not go to Moscow directly. With an FBI OK, Simpson shipped them to the Max Planck Institute at Lindau, and from there they were transshipped to IKI in Moscow. Sagdeyev gave them an English acronym DUCMA for dust particle counter and mass analyzer. To avoid any chance of international embarrassment he listed them as coming from Lindau until the Vegas were on their way to Venus.

The designs for the Vegas were just about complete when word came from the Soviet Academy of Sciences that in addition to the Halley experiments, the Vegas would carry landers to be dropped off into the atmosphere of Venus during the flyby.

This was the project of Jacques Blamont of the French space agency, Centre National d'Etudes Spatiales (CNES). Blamont had been working on this idea since the 1970s, and now it had finally been approved. In addition to all the Halley experiments, each Vega egg cup now had to hold an additional 3,300-pound egg, the Venus landers themselves carrying 10 experiments.

The Vegas would each release a balloon to which instruments were attached that would drift through the sulfuric acid atmosphere of Venus, their drift tracked by radio. Blamont lined up tracking stations from all over the world from Ulan Ude, Siberia, to Goldstone, California. NASA offered the services of the three 200-foot dishes of its deep space network, and these were joined by the world's largest radio receiver, that at Arecibo, Puerto Rico. While they floated in the Venusian atmosphere, the balloons could be located within 20 miles.

All this was wonderful and exciting science, but it created havoc at the IKI.

"We are always sentenced to Venus," Sagdeyev remarked. He and his engineers now began the frenzied scramble that Halley's relentless flight always precipitated when there were last-minute changes in spacecraft design.

It seems to be the usual practice with satellites that the instruments are constructed in laboratories in one place, then carefully transported to another where the engineers bind them into a unified body, *e pluribus unum*. A machine, yes, but an organism waiting to be animated.

The engineers at IKI received their 14 experiments before they redesigned the satellite for the Venus balloons. They still had to fit the landers and the experiments into the same space that had been designated for the experiments alone. An edge of one instrument case had to be filed to make it fit. A plate had to be wedged in to make it tight, the wiring rerouted and coupled again then tested. And all this time Halley was moving closer.

When they were finally tested in the clean room of the IKI in Moscow, all the instruments worked together like friendly partners. The satellites then were gingerly transported across the USSR to the Republic of Kazakhstan. It's desert country south of the Ural Mountains, north of Turkey, really a foreign country to Sagdeyev and the scientists and engineers who accompanied the Vegas from Moscow. The Kazakh speak a Turkish dialect, not Russian.

The modern Balkonur Space Center is located in the midst of this ancient desert. The enormous vehicle assembly building, large enough to accommodate the towering Proton rockets, has taken the place of reed huts.

Here the Vegas were integrated with the Protons, and the teams made their launch dates, December 15, 1984, for Vega I, December 21 for Vega II.

Venus, their first target, was 86½ million miles across space, but the Vegas weren't aimed directly at the planet. Instead, for a while the Vegas first circled the Sun following the orbit of the Earth, taking advantage of its gravitational field as Giotto did, using the force of gravity, and saving precious propellant.

At a prefigured moment, mission control fired thrusters, which moved the Vegas out of the Earth's orbit toward Venus. This flight plan was not as complicated as the gyrations that Farquhar had to put ICE through, but still the calculations for it kept the computers busy for a week.

During the 6-month trip to Venus the data the spacecraft gathered was stored on a tape recorder and transmitted every 20 days not only to Moscow and the Eastern European members of Intercosmos, but also to JPL, and to the stations of NASA's deep space network.

The Vegas arrived at Venus four days apart, June 11 and 15, 1985. The mission's navigational officers had been given radar maps of Venus made by Pioneer 12, which they used to tell them exactly where to release the landers.

When the landers were 33 miles above the surface of the planet, the balloons were released and their radios started bleeping to the tracking network Blamont set up for CNES.

Then the thrusters were fired so the Vegas would circle Venus and, using the boost gravity gave them here, the spacecraft made another loop of the Sun, their flight plotted so that in eight and a half months the Vegas would rendezvous with Comet Halley. During this leg of the flight, payload officers activated Simpson's dust analyzer to monitor the interplanetary dust (the data were transmitted to Simpson in Chicago), and other instruments were turned on to monitor the solar wind.

Sagdeyev and his team were excited over the way the Vegas

were functioning. Recalling the tumultuous days and frenzied nights without sleep, they felt that all their efforts had been well worth it. So far.

Sagdeyev was still worried about the winged solar panels. When the spacecraft dove into the coma of Halley and the panels were hit by the rifling dust particles, how much power would be lost? Would the batteries cover the loss? Would the ions in the comet's atmosphere form a cloud at the front of the spacecraft and interfere with communication? Or might they even get through the shields and interfere with the function of the instruments?

The newspapers thought they had a good story with renaming Giotto kamikaze. Now they extended their speculation, saying that the comet would destroy any spacecraft that tried to penetrate it searching for its secrets. Comet Halley would destroy the Vegas, and Giotto, and Giacobini-Zinner would certainly do in ICE.

Bonnet, Reinhard, and Farquhar decided on their responses in advance. It was a no-lose proposition. If the comets destroyed or crippled spacecraft, they would announce bravely, "We gave it our best shot. Comets are truly formidable cosmological objects." If the spacecraft survived, there would be cheering from Moscow, from Darmstadt, and from Goddard. Sagdeyev believed his own Vegas would survive. Farquhar and Reinhard thought they had a 50-50 chance.

12

Three Birds Flying

Kamikaze was not the Japanese plan, at least not in the early 1980s when the Institute of Space and Astronautical Science (ISAS) was preparing the first of its twin Halley satellites, Sakigake (MS-T5), meaning "pioneer." Sakigake was targeted to observe Comet Halley from a safe distance of 4 million miles. Suisei (meaning "Comet"), the second satellite, to be launched eight months later, would fly to 120,000 miles from the nucleus. Both satellites would be safely out of target range of any rifling dust particles from the comet. Both would rendezvous between the comet and the Sun in mid-March 1986 at about the same time as the Vegas and Giotto would be penetrating the coma. At that time Comet Halley, having circled the Sun, would be on the other side of the solar system. The tail of a comet always points away from the Sun, of course, and on that side of the solar system the solar wind would drive the dust particles away from Sakigake and Suisei.

When in 1980, ESA had decided to go to Comet Halley on

its own, ISAS had already decided on a Halley mission. The new Japanese rocket, M-3SII, was a fifth generation Mu, 93 feet tall, 70 tons, three stages with boosters attached. In capabilities, it was between the Mercury-Redstone that took Alan Shepherd above Canaveral and the Mercury-Atlas that took John Glenn into orbit.

Mathematicians at the main center for ISAS, Komaba, near Tokyo, did some computer modeling and found that the M-3SII would lift a 400-pound payload into an orbit around the Sun. This set the weight limit for the spacecraft that could be constructed. There was also a space limit. The spacecraft had to fit inside the cone of the rocket, and so couldn't be more than 5 feet in diameter by 8 feet and a little more tall, including antennae. The ISAS scientists working within those parameters would construct a spacecraft to Halley that would be the smallest of the fleet, a baby next to the 1-ton Giotto and Spartan-Halley; almost invisible next to the 4-ton Vegas.

However, ISAS was moving into the wilderness of space, and you don't go into the wilderness blindly. The pioneer, Sakigake, to be launched 8 months earlier than Suisei, would scout ahead. If glitches cropped up in Sakigake, they could be removed from Suisei.

The two space buses, the carriers, were exactly the same. They were drum-shaped like Giotto, and like Giotto the solar cells were wrapped around them. To leave as much weight as possible for instruments and experiments, the designers used the lightest materials: carbon-reinforced plastic, aluminum alloy, titanium, and even Kevlar, which had been used on the second dust shield of Giotto.

The designers cut down the navigational system to the absolute minimum. Two hydrazine thrusters were mounted on opposite sides of the craft, but with only 44 pounds of fuel aboard. The mathematicians at Komaba worked out trajectory calculations. There could be no parking orbits, no circling the Earth and then a jump into the heliocentric orbit. The Mu rocket would have to take the satellites directly into the orbit around the Sun and directly into a flight path that would cross the orbit of Halley.

Like Giotto, Sakigake and Suisei would revolve as they cruised through space, which helped to stabilize them. Unlike Giotto, though, their rate of spin could be slowed only by firing some of the scarce hydrazine fuel.

This careful weight watching left the scientists room for only three experiments on Sakigake: these would measure the ions in the solar wind, their density, temperature, velocity, and their magnetic field.

The instruments, constructed by private firms, had to be tested before they flew. The best testing equipment was at Tsukuba, 48 miles northeast of Tokyo. Tsukuba is a city specially built for 7600 scientists so they can conduct their research and experiments and make their discoveries right where they live. Many of the scientists working there complain of being taken from their delicate homes with gardens into a kind of sterile, cement block incarceration. Nevertheless, the city has become a successful incubator for new scientific advances.

Tsukuba looks the part of a twenty-first century science-fiction city, something out of Woody Allen's *Sleeper*, with rows of supermodern apartment buildings. Where they end, laboratories and testing towers continue. The testing facilities include Japan's finest and largest thermal and vacuum chambers and a shake table capable of handling a 13-ton object.

It was these that ISAS wanted to use, the shake table and the vacuum for the instruments of Sakigake and later for Suisei. A bureaucractic clash resulted. These testing facilities are administered by the National Space and Development Agency (NASDA), an agency formed to develop commercial, not scientific, satellites.

"You may use our vacuum chamber and vibration tester," NASDA said, "but will you pay for the electric power and the liquid nitrogen coolant?" The ISAS answer was the institutional answer heard in all languages:

"It's not in the budget."

The dispute was finally resolved in the prime minister's cabinet, where the top echelon shuffled a budget item between the Ministry of International Trade and Industry, in which NASDA was located, and the Ministry of Education, Science, and Culture, in which ISAS was located.

The instruments easily passed their vacuum, temperature, and vibration tests and were subsequently integrated into the carrier at the ISAS clean room in Tokyo.

Now a fully operating spacecraft, Sakigake then was shipped the 600 miles to the Kagoshima Space Center. There, in the vehicle assembly tower, it was coupled with the third stage of the rocket. Two platforms, one on each side, took the engineers up and down the 30 feet of the rocket's height so they could connect the electical systems and enclose Sakigake and the third stage in the nose cone.

Even with launches restricted to 4 months out of the year, the fisherman of Ohsumi Peninsula complained that their catches were down. But as far as ISAS was concerned, January 8, 1985, was the day of a great launch. The M-3SII rocket successfully lifted Sakigake into its orbit around the Sun. Mission Control at Komaba used some of its valuable propellant to make some corrections to the orbit the first week, then checked the craft's position every 10 days.

Tracking the spacecraft, ISAS tested its network, especially their new Usuda Deep Space Center. The dish there is 200 inches wide, the equal of the dishes in the NASA network, which joined in the tracking.

Sakigake was spinning on its axis at 120 revolutions per minute. The mission controllers at Komaba slowed the spin to 30 revolutions per minute and then finally to 6 revolutions per minute. In March they successfully activated the experiments to monitor the solar wind.

Sakigake, the pioneer in the wilderness, had prepared the way for Suisei. Launched on August 18, 1985, Suisei made a straight path into its orbit around the Sun. No corrections necessary. The fan antenna stopped spinning automatically so communication could be established between the craft and mission contol. As he had done on Sakigake, the flight director slowed the spin of Suisei so the principal investigators could test the two experiments on board.

The first experiment monitored the interaction of solar wind with the ions of the atmosphere of Comet Halley. The second used an ultraviolet spectrometer, the most sophisticated of the instruments either of the twins carried. Images of light were amplified, converted by a CCD, then stored in a memory until they were transmitted to Earth whenever the mission controllers requested them.

The spectrometer was designed to make a series of images of the hydrogen corona that extends around any comet. For example, the corona around Kohoutek, which Carr, Gibson, and Pogue had recorded from Skylab, extended in a shell larger in diameter than the Sun.

Suisei had the advantage staying a safe distance away from Comet Halley. Mountain climbers far away see the whole mountain: those close up see details. Spacecraft close in to Halley wouldn't pick up the whole corona, but Suisei at 90,000 miles would record the activities occurring within it—bursting jets and holes created as the nucleus of Halley itself revolves and turns different surfaces to the Sun.

To test the imager, the flight director had to slow Suisei's spin to one revolution every 5 minutes. Then the imager, through its microprocessor, was ordered to search for the Earth. It found the Earth.

"Observe it." The imager carried out that order.

"Image." Mission Control received images of the Earth in the prescribed ultraviolet light waves and transmitted them on order. The test was a success.

The principal investigator of the imager, E. Kaneda of the Geophysical Laboratory, University of Tokyo, had another mission in mind too. He hoped to be able to image Giacobini-Zinner as ICE dove through its coma on September, 11, 1985, now only a few days away. Not only at Tokyo but all over the world, at Moscow, at Darmstadt, at JPL, and not least at Goddard Flight Center, scientists, principal investigators, and flight directors anxiously awaited word on what Giacobini-Zinner would do to ICE. What happened to ICE as it crashed through Giacobini-Zinner's dust storms, ion storms, and magnetic fields could happen to Giotto and the Vegas when they entered the coma of Halley. There were construction parallels in all the comet probes. If the solar cells of ICE were ripped up or knocked out or both, the team at Darmstadt could see that happening to Giotto. And if ICE's wrapped solar cells were vulnerable, Sagdeyev could see his precious Vega's windmill blades turn into twisted garbage.

The antennae of ICE were especially vulnerable. There were seven of them sticking out on all sides of the barrel as if ICE had been the target of encircling spearmen. Dust particles might force the spears to bend around the barrel or even snap them off. Reinhard didn't believe the radio dish of Giotto would be as vulnerable, but Vega had antennae like television aerials on the roofs of apartment houses. The television aerials could easily be twisted into silence.

One of the most extensive tracking efforts in the history of satellites kept contact with ICE throughout its 21-month trip across space to Giacobini-Zinner. No TV coverage, nothing visual. The contact was electrical, with signals coming down to computer monitors with rolling lines of figures across the terminals, creating graphs of dancing peaks and valleys. NASA's deep space network followed ICE's progress with its chain of 200-foot dish receivers. At Arecibo, Puerto Rico, the huge 990-foot radio dish listened for every tiny beep. In central Japan, the Usuda tracking station also followed the flying satellite.

The radio system of ICE was originally designed to transmit only for about 9 million miles from its libration point. Now the signal eventually had to carry for over 43 million miles. The tracking stations not only increased the power of their systems to cover the extra mileage but also adapted their receivers for the wavelengths that ICE transmitted. At JPL, Yeomans and

Brandt watched the figures relayed from Goldstone, the network station in the Mojave Desert, while Farquhar and his associates watched at Goddard.

The comet had brightened up after its recovery by Kitt Peak. Big telescopes with their CCDs were able to fix on it, then medium telescopes like Emerson's Schmidt, and finally bright-eyed individual comet chasers with large binoculars. Sightings were sent to Steve Edberg at the International Halley Watch in Pasadena, and he relayed carefully calculated positions to Yeomans and Farquhar.

But Comet Giacobini-Zinner was scudding, leaping, and jerking along its orbit in a private dance. Yeomans attributed the dancing to the "youthful exuberance" of a young comet. The jets of dust and gases the solar wind and the Sun's radiance shot from the comet's nucleus were especially forceful.

"Just when you think you know where it is," Yeomans said, "it thumbs its nose at you. 'All right, fella, try again,' it tells you."

On June 6, 1985, when the Vegas were about a week's journey away from Venus, the erratic Giacobini-Zinner was about crossing the orbit of Mars. Yeomans and Brandt looked at the figures they had on ICE and called Farquhar.

"The bird's going to miss its nest by about 124,000 miles," they told him. So Farquhar went to work shifting the aim. From 8 A.M. until 12:30 P.M. he shot 2000 pulses out of the thrusters, blessing the day he had put eight tanks of hydrazine aboard. Farquhar re-aimed ICE to cross Giacobini-Zinner's path 16,200 miles from the nucleus of the comet. ICE continued on its new path at 45,000 miles per hour, crossing about a million miles of space a day.

On August 30, when ICE was 12 days away from its dive, astronomers at Goddard turned the International Ultraviolet Explorer (IUE), which was on orbit, to observe Giacobini-Zinner. The IUE's primary telescope is 12 feet long with an 18-inch mirror. The cooperative project of NASA, ESA, and the British Research Council, IUE was launched in January 1978 and was supposed to operate for 5 years at best. It's still working beautifully. Among the instrument's accomplishments is being able to locate an object in space within one arcsecond—0.36 degree.

The IUE gave Farquhar an accurate fix on the dancing comet that was Giacobini-Zinner. Farquhar punched out some more hydrazine, this time only 229 shots, and re-aimed ICE within 6000 miles of the comet's nucleus.

When ICE was only a few days away from the comet, the signals it was sending into the flight control building at Goddard were coming in garbled and incomprehensible. Farquhar was worried. He knew it would not be long before the scientists who had experiments on board ICE, principal investigators from all over the world, would arrive wanting to see their data. Of the 13 experiments, seven were operating completely, the rest only partially. But these latter should be able to report some data. Reporters from TV and other media were already filling up the motels between the town of Greenbelt, Maryland, where Goddard is located, and Washington.

The team at Goddard couldn't understand why it was receiving such garbled data. Goldstone was getting a good signal. So was JPL. Then somebody discovered that a crew repairing the roof of the flight control building had surrounded the antenna with all sorts of junk, including a pot of boiling tar. They were politely persuaded to repair another Goddard roof, and when they left, the signals began arriving downstairs in perfect order.

Scientists watched their monitors. Cursors squirted across screens, seeding the figures behind them. ICE was 44 million miles away, and it took the figures four minutes to arrive at the computers.

ICE was about 600,000 miles from its target, crossing an area of transition in the ocean of space, a stream of relatively undisturbed plasma. The ionization of the comet's gases can be a slow process, and some ions are created at great distances from the comet. The instruments on ICE sensed these distant ions floating along the solar wind.

ICE was less than a day from Giacobini-Zinner. Everything had come together, the whirling around the Moon, the tracking networks, the pulsing hydrazine. Now two time lines were meeting, Giacobini-Zinner from four billion years ago, ICE from the twentieth century.

E. Kaneda, at the University of Tokyo, focused the imagers of Suisei and Sakigake on Giacobini-Zinner hoping to register the passage of ICE through the comet's coma. It was unfortunate that the images transmitted were at first very faint. But later, before Suisei and Sakigake continued in their paths toward Comet Halley, mission controllers at the Komada tracking station did receive some good images of the comet—though none of ICE.

Eleven and a half hours later, September 11, 1985, ICE sailed into the waves the solar wind formed around the bow of Giacobini-Zinner. This bow wave had the same shape and impact as water breaking around the bow of a large ship, which will rock a small boat some distance away. The scientists called these waves the ion sheath. Instruments at Goddard showed that temperatures here shot up and down, and the speed of the solar wind was slow, then fast, then slow again. The intensity of the magnetic field varied wildly.

The turbulence stopped as soon as ICE entered the envelope of flying dust particles. Now, every second or so the antennae recorded a hit. The data still arrived on the screens 4 minutes after the events occurred. The scientists wondered anxiously if ICE would suddenly stop signaling and the screens go blank. Would the dust inflict the damage that was predicted?

The scientists kept their eyes fixed on the computer monitors. Figures continued running down the screen. There was no loss of power, no calamity, no damage. ICE sailed through the rifling dust particles unscathed. Then ICE plunged into the coma of Giacobini-Zinner at 6:50 A.M. (EDT), September 11,

1985. At last, for the first time in history, humankind had reached into a comet. Some of the scientists clapped. Still others cheered on the spacecraft vocally as if, across the cosmos, it could be encouraged like an Olympic runner.

About 100 million miles away, the Pioneer Venus orbiter was circling and scanning the planet Venus with its ultraviolet spectrometer. Now its controllers at NASA's Ames Research Center, Mountain View, California, turned the satellite away from the planet so its spectrometer could scan Giacobini-Zinner for any unusual activity that might result from the penetration of ICE into the comet's coma.

At the space science building at the University of Colorado, the principal investigator of the ultraviolet experiment, Ian Stewart (an Irish mathematician and astrophysicist at LASP) carefully followed signals Pioneer Venus sent down to the computer consoles. According to the figures Stewart received, Giacobini-Zinner ignored ICE, this brash, twentieth-century invader.

ICE was now 9700 miles from the comet's nucleus. Here in the coma, the plasma tail was being formed. Figures on the computers indicated that in this region a strong magnetic field flowed away from the Sun, confirming in fact what before had only been theorized. Then ICE crossed through a neutral region, the neutral sheet, they called it, borrowing from the geologists, a transition layer.

On the other side of the neutral sheet ICE, like Alice, went through a looking glass. As it penetrated this other side of the tail, ICE entered a mirror world of what it had experienced earlier. Scientists discovered that here the magnetic field flowed in the opposite direction, now toward the Sun. The spacecraft soared within 4900 miles of the nucleus, much closer than Farquhar and the principal investigators had intended.

ICE finally left the coma, and sailed harmlessly through the envelope of dust on the other side of the comet, and then through the ion-sheath a second time.

At 8:22 figures on the computer screens at Goddard told

the scientists ICE had once again entered into the more subdued stream of faraway ions in the solar wind. In 20 minutes ICE finished its historic rendezvous, its 15,000-mile trip through Giacobini-Zinner, humankind's first physical encounter with a comet in flight.

Three days later, Giacobini-Zinner, soaring away from the Sun, would pass within 2 degrees of Comet Halley, which was on its approach to the Sun. However, Halley was still three months away. On Mt. Thorodin, outside Denver, Gary Emerson, now well known for his earlier photographs of Comet Kohoutek, was to get a rare shot of two comets on one plate of his Schmidt camera, as would many astrophotographers who knew about the close encounter.

Back at Goddard, the astronomers leaned back and relaxed. Their comet mission was over, but their computer printouts would give them plenty to study in the future. All the work and effort that had gone into the transformation of ISEE-3 into ICE and its diversion across the 44 million miles of space was worth it. They would be able to compare their findings about Giacobini-Zinner with what they would learn about Halley.

The word went out to Darmstadt and Moscow. ICE had sailed through Giacobini-Zinner intact, with no bent antennae, no ripped solar cells. It was now on its way to encounter Halley in March. Sagdeyev had flown all the way from Moscow to Goddard to follow the mission and he congratulated Farquhar personally.

However, although ICE had not been damaged that was no automatic guarantee that the Vegas and Giotto would survive Halley's coma. Brandt and many other cometary astronomers believed that the dust of Comet Halley could be 100 times denser than that of Giacobini-Zinner. And then, too, everyone kept in mind Whipple's immortal proverb: "Never trust a comet."

13

Preparing the Busload

During the first two months of 1985 the Vegas were orbiting the Earth, ready to be whipped toward Venus. Sakigake had left Kagoshima and was gliding toward its inspection point, 4,200,000 miles from Comet Halley. At Toulouse, France, the electrical circuits of Giotto were being grafted into those of the experiments and instruments.

At LASP in Boulder, Colorado, Stern had prepared his computer system so that the shuttle's mission specialist could update Spartan-Halley's microprocessor when the spacecraft would be on orbit. But NASA still hadn't assigned Spartan-Halley to a crew or a crew to a shuttle. Meanwhile, Barth told Stern to prepare the cameras.

Ever since the first project initiation conference at Goddard, Stern had been bubbling with his dream of getting the first pictures of Halley from space, of beating out the already departed Vegas, and the ready-to-fly Giotto.

The Spartan-Halley cameras had to take pictures of the

comet when Halley was within an angle of 6 degrees from the Sun. The comet would be half as close to the Sun as the Earth. Here Comet Halley would be subjected to the full force of billions of horsepower from the Sun's nuclear radiation and millions of tons of force from the solar wind. The photographs Spartan-Halley took would provide a record of comet convulsions never before seen. The Johnson Space Center lent the Spartan mission a 35-mm Nikon they usually used for astronaut training, plus two 135-mm telescopic lenses.

It was a good camera but not much different from one that could be bought at any store. How would you take pictures of a comet 220 million miles away in the black sky of space? The Skylab crew had used the camera but not as close to the Sun as Spartan would be.

Stern thought if anyone in the Denver area could help solve the camera problem, it would be Gary Emerson at the E. E. Barnard Observatory on Mt. Thorodin. Stern took many different kinds of films up to the observatory, and the two of them tested the films at different exposures shooting the North American nebula. It was a long process that took several all night stints. Paula Emerson kept the experimenters supplied with coffee. When the series of photos was complete, Stern could extrapolate data from these experiments to set up the cameras on Spartan.

Wilshusen, Leyner, and others at 55th Street were redesigning spectrometers, adapting 1985 circuitry and techniques to the 1976 instruments. The spectrometers were a great deal more complicated than the one they had used on the UVA 84 in Alaska. That spectrometer had analyzed only one wavelength. The new Spartan-Halley had an elaborate detector, a Codacon, that actually coded 1024 different ultraviolet wavelengths so each could be recorded on tape for later analysis.

The telescopes of the spectrometers were baffled inside, but as close as the comet would be to the Sun, this baffling would not provide enough shading for the delicate instrument to operate. To compensate for the excess light, the designs of the carrier now called for a sunshade, a projection angled off one side of the spacecraft like an overlong visor on a cap. This sunshade would shadow both the spectrometers and the cameras.

Barth had gotten the LASP Halley mission moved up on the Spartan list of the carriers waiting for a shuttle ride. Spartan-102/Halley was now number two. Number one, the first Spartan, was a Navy Research Laboratory x-ray experiment supposedly flying on *Discovery*. *Discovery* finally made its maiden flight August 31, 1984, but did so without Spartan 1, which had been replaced by Navy and AT&T communication satellites plus a couple of secret projects. Spartan 1's bounce was discussed on a conference call that tied together Len Arnowitz, Morgan Windsor at Goddard, and Barth, Leyner, and Wilshusen at Boulder.

LASP: When will Spartan 1 fly?
Arnowitz: June '85 looks like.
LASP: That doesn't allow time enough to get the x-ray experiment off and Halley on.
Arnowitz: Yeah.
LASP (Desperately): What's the alternative? Redesign the Halley experiment for a rocket?

Arnowitz had another alternative. He assigned Spartan-102/Halley a new number: Spartan-203/Halley. The 2 meant Spartan-Halley had a new space bus, Spartan 2.

In one way Jones, Leyner, and Wilshusen were both happy about the new Spartan 2. The special payload division had made a lot of improvements in the carrier. Now the Halley mission could carry more propellant for control and maneuvering. The tape recorder was a 14-tracker and could carry a lot more data.

On the debit side, instruments had to be relocated and the wiring rerouted for the new bus. Jones put up a new schedule on his office wall. Everything had to be ready for integration at Goddard by July 4, 1985.

Although LASP hadn't heard officially, a rumor reached it that Spartan-Halley had been assigned to a crew and shuttle.

Stern, using his contacts from his earlier experience at the Johnson Space Center telephoned the various offices and learned the crew assigned to Spartan-Halley was probably 51-L, that commanded by Major Dick Scobee. Scobee had been second in command and pilot on *Challenger* for the "Solar Max" rescue. A satellite, the Solar Maximum Mission, had been launched in 1980 to study the Sun. Its electronic system failed in November 1980. On a special mission in 1984, *Challenger* brought Solar Max into the cargo bay using the shuttle robot arm. There, members of the crew repaired its electonics and then the robot arm lifted it back into orbit, thereby saving a $200 million satellite for five more years of active life.

One of the other four members of the 51-L crew included Lieutenant Colonel Ellie Onizuka. Onizuka was the University of Colorado's own, a graduate who made it big. He was their personal astronaut. Everybody on the Spartan project hoped Onizuka would be their mission specialist, but that assignment was up to the mission commander, Scobee. It wouldn't be Onizuka, Scobee explained. Onizuka had enough to do. He was mission specialist for a communications satellite, the other payload on 51-L. It was a very important communications satellite to be put in place as part of the tracking and relay satellite system (TDRS), one satellite of which was already on orbit.

The mission specialist for Spartan-Halley would be Ron McNair, a Ph.D. in physics from MIT. Scobee's assistant and pilot was Commander Mike Smith, one of the Navy's best test pilots. This was his first flight, but his record in training was so outstanding that he was slated to be a shuttle commander on his next assignment. This was the first time NASA had made an astronaut a flight commander before he'd even flown a mission.

The fifth member of the shuttle crew was Judy Resnik, a Ph.D. in electrical engineering from the University of Maryland. She didn't often talk about herself, but once as she was making suggestions about the wiring of Spartan, she added, as if she ought to explain her expertise, "On my income tax form I don't put 'astronaut,' I put 'engineer.' "

Resnik had the exacting job of manipulating the robot arm which was to deploy Spartan-Halley into space. It is the most complicated robot arm ever built, 50 feet long, and articulated at its shoulder, elbow, wrist, and fingers; each movement requires very delicate handling.

If for some reason Resnik could not deploy the satellite, Scobee had to maneuver the shuttle so that when the cargo bay opened, Spartan's two spectrometers were pointed in the right direction and could operate while still fastened in the bay. This plan was called the "In-the-Bay" mode.

Scobee's crew was assigned to *Challenger*. This shuttle was the second one built, and this flight would be its ninth. The five astronauts came up to Boulder several times to learn all they could about Spartan-Halley. LASP was not all that well known, and the astronauts did not understand what to expect. Among the astronauts, Onizuka had a reputation as a practical joker, and the crew was glad to have this chance to get back at him and needle him about coming from a mountain college.

Driving the 30 miles from Stapleton airport in Denver to Boulder the car passed a dilapidated frame of a farm house, long deserted.

"There it is, Ellie. Colorado University, huh? That's it." Onizuka just grinned.

After the astronauts' visit, Stern went down to the Johnson Space Center outside Houston to see Scobee and explain the high-low relay system he had devised for entering data into Spartan's navigational microprocessor when the spacecraft was being prepared to be deployed. Scobee was worried about adding more work to an astronaut's already strenuous duties.

"How long do you think the whole process'll take?" he asked.

"If you take your time, get it right, pay attention, it'll take half an hour," Stern answered. Scobee agreed to the plan and asked Onizuka to act as McNair's backup.

"It'll sure be easy to screw up the data and punch the wrong stuff in," McNair said when Stern laid it out for him. "And if it screws up, you know what they'll say," McNair continued. "They'll say Ronald McNair screwed up the mission." Stern, always optimistic, assured McNair he wouldn't screw it up, but McNair kept on questioning Stern until the astronaut was sure he understood just how the relays that Stern had set up were expected to work.

As often as possible, Stern intruded on McNair's and Onizuka's already crammed workdays and even crammed work nights to rehearse them in how to punch H's and L's.

The five members of the crew shared an office on the third floor of Building Four. It was a modern office, grey steel files and desks, one desk shoved up against the other, crowded conditions to get a crew used to up-close family togetherness. After all, the flight deck and middeck of a shuttle are even more cramped. The windows, though, had a pleasant view of one of the grass and tree malls that separate the buildings of the space center.

On the morning of June 17 the LASP crew gathered around TV sets at 55th Street to watch the launch of *Discovery* 51-G with Spartan 1 aboard. The launch went off smoothly. The engineers laughed and clapped, one hit another on the upper arm harder than he meant to. Both grinned. "Give me five," said another, and they slapped their palms. In six months it would be Spartan-Halley up there. All of them bolted back to work in a state of euphoria.

The *Discovery* flight was covered in detail by the three major American networks and Worldwide Television News with even a special broadcast to the Middle East. The coverage was not because of Spartan, but because one of the payload specialists was Prince Sultan ibn Salaman ibn Abd al Aziz Al Saud of Saudi Arabia, a graduate in petroleum engineering from the University of Denver.

Two days later aboard *Discovery*, Astronaut Shannon Lucid lifted Spartan 1 from the cargo bay with the shuttle's manipulator arm and deposited the carrier in space. The Spartan recorded x-ray readings from the center of the Milky Way galaxy, then a number of galaxies in the Perseus constellation. When the scan was over, Spartan's propellant was exhausted, and the spacecraft started to roll.

The commander of *Discovery*, Daniel Brandenstein, knew Lucid couldn't pick up a tumbling object and proceeded to roll the shuttle around Spartan at the identical speed, a maneuver called "rate matching." Rate matching made it seem as if Spartan was still. Lucid moved the robot arm precisely so it took hold of the grapple and moved Spartan gently back into its nest.

Except for running out of hydrazine the flight of Spartan 1 was successful. LASP and Goddard both were elated. So was Goddard. Their space bus worked.

Comet Halley was already halfway through the asteroid belt and headed toward the orbit of Mars; Earth was not far beyond. Even by July 1985, the Comet was still pretty far out with a magnitude of +15, weak for small telescopes. However, while Halley was as yet very faint, the furor over its impending arrival in the skies of Earth had gotten very loud. A once-in-a-lifetime spectacular was about to occur, and public excitement was building rapidly. Alan Shepard, our first astronaut in space, was host of a televsion special, *The Return of Comet Halley*. He was assisted by John Brandt, who explained about the Comet's place of origin, Oort's Cloud, to their large audience.

Since the best "seeing" of the Comet would occur in the southern latitudes, travel agencies were organizing trips to Alice Springs, Australia, to Auckland, New Zealand (with Carl Sagan), to the Fiji Islands, actually to almost anywhere in the Southern Hemisphere. Comet Halley cruises were going to Rio de Janiero, up the Amazon, through the Panama Canal, westward to the East (to Bangkok and to Hong Kong), and eastward to the West (to Zaire, Nigeria, and Gabon).

All over this country, but especially in the southern United States, people walked out at night looking for the comet. Few of them had any idea where in the sky to look. McDonald Observatory, in the Davis Mountains of southern Texas, handled a lot of inquiries and kept a graduate student or an astronomer constantly at the ready on a telephone switchboard.

"Can you tell me when and where I can see the comet?"

"Where are you located?" switchboard asked.

"Virginia."

"Got a latitude and longitude?"

"No." Whoever was on the board would have to get the name of the town and look it up in an atlas, then work from there.

Or the observatory would get a mournful cry.

"We were out last night looking for the comet but couldn't find it. We found the Big Dipper, though."

"Look the other way," McDonald advised.

A service in Atlantic City promised to help the searchers out. They would keep their subscribers informed of the comet's progress with a weekly news bulletin. For the would-be observer who also had a home computer, companies advertised software programs that would track the comet on a day-by-day basis.

The scholar could order a comet library, five books worth $77.35, now only $3.95. Robert Chapman and John Brandt had published one of the standard texts on comets, which they now revised for a popular audience. However, these experts had to compete with many other books. Most book stores had a section labeled "Halley's Comet," which was usually dominated by the best selling Carl Sagan-Ann Druyan, *Comet.* Nigel Calder's comet book was still on the Halley-Comet shelf. Two years earlier Patrick Moore had addressed the Halley's Comet Society in London. He had revised an earlier book into *The Return of Halley's Comet.*

The original "Once in a Lifetime" T-shirt now had all kinds of competition. One that came in tiny sizes promised, "I'll see you in 2061." A huge poster entitled "The Return of Halley's Comet" sold better than the T-shirts: it traced the path of the Comet across the sky from January 1985 through January 1987 and was decorated with reproductions of Giotto's mural, the 1066 appearance from the Bayeux tapestry, and the recovery as Danielson and Jewett had it on their printout.

All kinds of telescopes and binoculars were advertised as "comet finders." Imitating their predecessors 76 years earlier, a California vineyard promised to put some of its '86 vintage aside as "comet wine," and special champagne glasses etched with a flying comet were available to toast Halley.

Viewers of Louis Rukeyser's *Wall Street Week*, the PBS investment show, wondered if the companies that offered all these memorabilia wouldn't be a profitable source of investment. Rukeyser recommended two or three.

The warnings of the anti-Halley forces were lost in the media whirlwind: the pessimists reminding their few followers that Halley had brought William the Bastard to England, the Turks to Vienna, and death to Mark Twain, Leo Tolstoy, and Florence Nightingale. And if comets annihilated the dinosaurs, as many believed, could frail homo sapiens be safe?

In Boulder the LASP team could shut out the public frenzy. But frenzy or not, Comet Halley kept approaching at 73,000 miles per hour. The beginning of July was the absolutely final deadline when the Spartan instruments had to be completed, mounted on a metal plate (called the optical bench), and shipped to Goddard. Here they had to be integrated into the Spartan carrier, shipped to the Kennedy Space Center, and loaded to fly in time to make the January launch window.

Neither the spectrometers, the star tracker, or cameras had been mounted on the optical bench. Programming for the computer still needed to be finished. The electric cabling wasn't yet complete. And everything still had to be tested. Some of the work that should have been done in Boulder had to be completed at Goddard.

On July 5 Wilshusen, Kohnert, and Stern flew to Goddard. The others would follow; all of them knew they had a grueling schedule ahead of them. Only Wilshusen was confident they could finish in time to fly.

14

The Sunlamp Flight

In the clean-room at the Goddard Space Flight Center, Rick Kohnert was looking at the star tracker in front of him and yelling. Normally Kohnert is a quiet, soft-voiced engineer. His fellow engineers have to listen carefully when he speaks.

Since Kohnert is not the kind of person who yells, his howls brought other engineers at the Instrument Construction Building running right over. The engineers and the scientists with them had been integrating the Halley instruments into the Spartan carrier and preparing the unified spacecraft for shipment to Cape Canaveral, the Kennedy launch pad, and eventual flight.

"The door won't drop," Kohnert explained. "There's not enough clearance." In front of Kohnert was one of the prime navigational instruments, the star tracker, fixed inside its aluminum housing. Closed, the door of the tracker protects it from the glare of the Sun in space, which would otherwise burn it out. During the flight, when it's time for the tracker to locate

Canopus, the door rolls down from the tracker lens. Kohnert had just discovered there wasn't enough room between the instrument and its housing for the door to drop.

"I've got to tear out this whole goddamn thing and get more clearance," Kohnert said frantically. It could be a month's work, but they didn't have a month. They would be lucky if they could hold off the integration for a week. Three days might be a possibility. He'd have to do a month's work in three days.

As soon as Stern, Kohnert, and Wilshusen had arrived at Goddard just after July 4, 1985, they began the work of integration at "fast forward." Fast forward is psychological. Everything around you has to move at fast forward, too. Anything that slows down the surge produces an explosion like Kohnert's.

Wilshusen had warned about that.

"Don't be Mr. Speedy," he said. "You just make mistakes." Calmed down, Kohnert and a couple of others went to work dismantling the star tracker, shaving the housing, and reassembling everything.

Wilshusen worked over the electric circuitry, painstakingly gathering the wires and wrapping them into cables, ropes of wire circuits called "harnesses."

"We'll just have to stretch the days," Wilshusen added. As they had at Boulder, the days here stretched: 9 hours went to 10, then to 12. The teams worked a half-day every Saturday, then all day, then all day Sundays. They didn't know Thursday from Friday from Sunday because it didn't matter.

But whatever happened seemed calculated to move fast forward into slow motion. The instruments were also fitted in cases like the star tracker. The case for one of the spectrometers hadn't been finished nor had it even arrived from Boulder. Neither had the rubber baby buggy bumpers. Spectrometers have doors that flop open at the beginning of the mission; the flop is halted by these bumpers, which continued to be missing.

When the camera boxes arrived, they were the wrong size. The boxes went back to Boulder. Rebuilt boxes of the right size were sent to Goddard, but during their tests in a vacuum chamber to see if they would survive the vacuum of space they failed. The windows blew out of the boxes.

"Don't send the boxes back to Boulder again," Kohnert said. "We won't see them until 1988. Have them send down a set of windows. I'm sure I can seal them in."

The Navy did its share to keep the workload set in fast forward. Palmer was called to active duty, and the programming was left unfinished. As the new programmer, Barth hired Ross Jacobsen, who looked like the classic computer programmer—tall, lanky, with well worn levis and tennis shoes, his shirttail out, and conveying a languorous demeanor.

The languor was an affectation, both physically and mentally. Jacobsen was a karate instructor at the University of Colorado; no languor here. And when you see him in his white *gi*, you realize he's not lanky, but broad-shouldered and rugged. And mentally he was judo sharp. He had a master's degree in electrical engineering and was the inventor of a machine that registers brain waves on computer printouts.

Jacobsen had to coordinate his programming with what Palmer left, then build on Palmer's bit by bit.

Spartan had to be ready to go into the *Challenger* bay, and whatever was done at Goddard had to conform to shuttle safety standards.

Stern was the liaison between Morgan Windsor, in charge of the integration at Goddard, and the flight operations departments at the Kennedy and Johnson space centers. In the vehicle assembly building of the Kennedy Space Center Spartan would be loaded into the shuttle cargo bay, and from there *Challenger* would be moved to the launch pad and fired into space. As soon as *Challenger* would clear the launch tower and was airborne, the Mission Control Center at the Johnson Space Center would monitor the rest of the flight.

Very detailed directions on how Spartan should be loaded, deployed, and retrieved for launch and flight had been issued by NASA. Stern often flew to Cape Canaveral and to Houston to pick these up: "Mission Operation Directives," "Operations

Criteria," "Procedure Development," 8½ × 11-inch booklets ranging in length from 10 to 50 pages. Windsor's office was piled high with them. Laudadio, the safety officer, checked to see that the directives were followed.

The flight director for 51-L, Randy Stone, and the flight controllers at Johnson were as enthusiastic about launching Spartan on a U.S. Halley mission as LASP and the Goddard teams were, but even before launch *Challenger* would go through about 2000 safety checks. The flight control personnel invited Windsor and members of both the LASP and Goddard teams down to Kennedy and Johnson to learn about these safety checks and about exactly what happens during both launching at Kennedy and flight control at Johnson.

The *Challenger* crew frequently made return visits to Goddard to see how the integration was getting along. McNair and Onizuka practiced the relay system every chance they got. Resnik worried about the possibility of Spartan-Halley tumbling around in space the way the first Spartan had. In space, the maneuvers of the satellite would be controlled by argon gas under pressure released through nozzles. Stern added enough argon to what Spartan-Halley already carried making sure there was a real margin for success, the Rule of Redundancy.

In addition to the argon redundancy, Jim Watzin, the young attitude control engineer at Goddard, worked out a magnetic control system to prevent Spartan from tumbling, a backup to the redundancy.

Resnik really liked the antitumbling device. The controls Watzin designed for the little spacecraft sensed the direction of the Earth's magnetic field and kept the spacecraft aligned to it. If Spartan-Halley started tumbling, special sensors would fire jets automatically to realign the spacecraft. And the system had its own battery and wiring system. Even if Spartan's main power system failed, the magnetic control system still operated—a backup of the backup system.

In the newspapers and on TV the Halley story continued to grow. Danielson appeared on a PBS special and told the story of the recovery. Alfred University, Alfred, New York, published pictures made on its 20-incher. Emerson was taking telescope photographs of Halley every clear night.

Interest in Comet Halley didn't need any artificial stimulation. In the fall of 1985 the smallest telescopes, the 4-inchers, even binoculars, were able to pick up the Comet appearing at a magnitude around plus eight between the constellations of Taurus and Aries.

At the University of Colorado, the Sommers-Bausch Observatory opened its telescopes to the public. Long before twilight when the door of the observatory would be opened, lines were already forming from the observatory door all the way down the narrow hill road. A 2-hour wait for the once-in-a-lifetime experience wasn't unusual.

In Texas, the McDonald Observatory public information office and the Astronomical Society, an amateur group, set up telescopes in King Falls State Park. So many people, somewhere between 4000 and 5000, drove into the Park that the Comet vanished from sight before some of them could get to a telescope. VIP's had to be entertained at the observatory itself, and one of the astronomers, Anita Cochran or Ed Barker, would guide them in sky searching.

The VIP would usually pick up one of the stars first.

"Oh, no," Cochran explained when she checked where the telescope was pointed. "That's a star. The Comet's much bigger. It's as big as the Moon." She could tell from his expression he didn't believe her. Then he would find the Comet in the telescope. And yes, now it was as big as the Moon.

At various places along the road out of Fort Davis, 30 miles from McDonald Observatory, people were camped each night in recreation vehicles, even tents, looking at Comet Halley. Sometimes others would drive right up the Mt. Locke road close to the Observatory. McDonald personnel politely discouraged this, because visitors wouldn't observe light discipline and used bright flashlights or even their headlights.

One night two people in an old Subaru drove up and set

up a small telescope for themselves. One of the McDonald astronomers went out to talk to them. They had driven all the way down from Rochester, New York.

"Why did you come here?" the astronomer asked. "There are plenty of other places closer to Rochester just as good."

"We heard McDonald has the darkest nights in the United States," they explained. The astronomer didn't want to contradict the McDonald publicity. Anyway, since the two men from Rochester seemed to know about light pollution, they were allowed to stay where they were. They observed the Comet just that one evening, then packed up their gear and drove back to Rochester.

Out of Denver and the surrounding towns, people trooped to Emerson's Barnard Observatory. A couple drove up from Boulder and brought with them a very old relative. He had been an amateur astronomer in 1910 and had seen the comet then on his brass refractor. At the same time he was at the observatory, a group of children arrived up the mountain from a preschool. Emerson knew at least one of all these children would see Comet Halley when it came around again in 2061.

"We had three apparitions of the Comet right here at the same time," Emerson said.

The LASP team at Goddard had to miss the appearance of the Comet right then. They were working until 11 or 12 P.M. and would stagger in again at 8 A.M. Even with 16-hour days, the integration was still behind schedule. Barth came over from Boulder and worked as an extra engineer. Wilshusen kept checking the harnesses and circuitry. The rubber baby buggy bumpers arrived finally and were installed on the spectrometer doors.

To fit the star tracker, spectroscopes, and cameras exactly, Kohnert had more drilling, filing, and tightening to do. Wilshusen checked the alignment through the theodolite, a very accurate survey telescope, fixing the instruments vertically and horizontally to 10 thousandths of an inch.

The internal organs of Spartan had survived the tests of the vacuum, the cold and the heat, as well as the shakes. The spacecraft now rested on a blue dolly, inert.

Will it fly, Orville?

We'll find out. We'll fool Spartan. We'll take it right through its mission, all 44 hours, right here at Goddard, but let Spartan-Halley believe it's in space, a make-believe flight inside the clean room.

The LASP and the Goddard teams wired 18 pieces of test equipment, test boards with vertical and horizontal rows of lights, switches, and dials, computer keyboards, printers, and screens that would display moving graph sequences, the lines as jagged as the outlines of a neurotic mountain range.

The test boards and computers would signal what the navigational controls and experiments would be doing as Spartan flew, theoretically, around the Earth. For the last time human beings would be able to watch what was going on. Spartan-Halley, two hundred miles above the Earth during its actual flight, would be on its own. Then Spartan-Halley would not be able to call for help. Only the Watzin and the Palmer-Jacobsen-Stern space brain would be in command. Only silicon chips would know where the spacecraft was, what it had to do.

Before Spartan came to life, all the gauges and screens registered zero, showing that everything aboard was truly asleep, no power escaping when and where it shouldn't. Spartan now believed the launch had taken place and that, after a 57-hour trip, it was 178 nautical miles in space.

McNair's APC calculator was tied electrically into the experiment microprocessor, the Captain Kirk computer. McNair punched the buttons that sent power into the craft. The little city was coming to life. A few board lights were on. Dial needles swung. Numbers and lines spilled out on the monitors of the test computers. The ion pump was go. The battery was go. The heat was on. After a 10-minute wait, the instruments were ready to operate.

McNair cleared the microprocessor and entered a simulated date and time. Stern set up the relay highs and lows for a January 15, 1986, time of day in minutes and seconds.

"Relay command number 12, low to high." McNair punched it in.

"Relay command number 13, low to high." OK. McNair punched in the shuttle's exact altitude. All the rehearsals paid off. No glitches. Now the microprocessor knew the position of the Sun and of Canopus for that date and exact time.

After running down a final checklist, McNair released his APC. Spartan was ready to be deployed. A switch signaled that Spartan was free in space. The Scotty computer asked the Kirk computer,

"What do we do?"

"Pirouette," ordered Kirk. Scotty fired the jets, or thought he did. No argon was aboard yet. Stern saw the firing turn on the correct lights and register on the computer terminal in front of him. He nodded and a quiet cheer went up all around.

Now the two computers in Spartan were working together. Camera doors opened. The electrical circuits on the cameras operated. The cameras would take pictures of the sky around Spartan and show exactly where the spacecraft was located. Camera doors closed.

That job completed, the Kirk computer figured the length of the orbit from the altitude of the shuttle. The microprocessors directed Spartan around the Earth on its first orbit, a lazy orbit for Spartan, but not for Stern and the Goddard mission monitors. As Spartan went into the shadow of the Earth, Watzin, Stern, and Windsor watched the temperature-drop register on a computer monitor.

As Spartan emerged into its day on the sunny side of the Earth, the first Sun sensors on the edge of the craft's sunshade picked up the brightest object in the sky, the Sun. Here in the clean-room, the sensors are locating a sunlamp. There are a series of three Sun sensors, and each pinpointed the electric Sun more accurately than the one before it for a precise fix.

The Scotty computer received the order to hunt for Canopus at an angle of 90 degrees. The door of the star tracker that Kohnert sweated over opened, and the tracker began its search for an object of the exact magnitude of Canopus. A series of lamps faked other stars. The Star tracker found one. No, that's not the right intensity. A second. No, not that. Only Canopus' magnitude registered perfectly on the chips Palmer and Jacobsen programmed. The tracker found it.

At high noon on Spartan's fourth orbit, using the information on the exact position of the Sun and Canopus, the Kirk computer looked up the position of Halley relative to these two celestial objects. Kirk ordered Scotty to fire jets and move Spartan horizontally and vertically until it was fixed on the position of the comet.

Now Spartan was ready to make four scans of Halley's nucleus, coma, and tail in 1024 spectral lines of ultraviolet light. Timers sprung open the spectrometer doors which, closed during launch, bumped against the rubber baby buggy bumpers. The ultraviolet light could enter the telescopes, bounce off mirrors, and by means of the Codacon be separated into wavelengths, which are then registered on the tape recorder. At the same time, the cameras photographed the comet.

The sunlamp's intensity was reduced to that of a false twilight. Then the globe of the Earth shaded much of Halley's coma but not the tail. The spectrometers received special scans of the tail, and Stern received *Life* magazine pictures.

Spartan believed it was circling the Earth for twenty-eight 90-minute orbits, a total of 44 hours, covering the time that scientists then believed Comet Halley took to make one revolution—a theoretical Comet Halley day.

Spartan ran its 28 orbits. On the 29th orbit, Spartan relaxed, lowered its power, and waited for Scobee to bring the *Challenger* over to it.

The crews at Goddard were watching the entire 44 hours in alternating 8-hour shifts. Now, tired but glad it was almost over, they applauded the lowering of power registering on the test equipment. McNair was ready with his APC to go through the postretrieval check.

Just at this point all the lights in the clean-room went out. Absolute black. Everything was down. The invisible fans of the clean-room were silent.

In all clean-rooms the air is continually filtered, the dust and other contaminants taken down to the floor and out of the room. People working in a clean-room are dressed in white bunny suits similar to those worn by surgeons from St. Elsewhere.

Germs, however, aren't the enemy. Earthly dust is. It produces shorts and high voltage arcs. Dust can attract molecules and cause all sorts of false readings.

With the fans down, Spartan was no longer in a clean-room. With only flashlights to guide them, members of the crew rushed and covered Spartan with a sterile plastic bag.

The test had almost ended. Windsor believed Spartan had been bagged before it was contaminated. They checked the printouts from the test flight to decide if it would be necessary to finish it up or even do it again. It wasn't.

A special simulation flight, the astronauts' final rehearsal for the deployment and retrieval of Spartan in the shuttle mockup, was scheduled to take place at Johnson in January. McNair would go through the retrieval then. There wasn't time at Goddard to go through another 44-hour run.

Just before Thanksgiving, Spartan-Halley was ready to be shipped to Canaveral and Kennedy. There it would become part of Shuttle Flight 51-L.

Wilshusen, Windsor, Kohnert, Stern, and the rest stood in silence looking at their little spacecraft being prepared for the sealed truck that would take it to Florida. It just didn't seem possible they had actually managed to get Spartan-Halley completed in time. They stared at the little craft with a variety of feelings: there should be one more connection to make, Wilshusen felt; Kohnert felt there should be one more door to attach; Jacobsen, a thousand more lines of software. But there was Spartan, with a potential for life, all ready to fly through space.

"Jeesus Keerist!" Someone said. That broke the spell. Everybody laughed.

"That bastard's going to fly."

"Abso-goddamn-lutely."

"Halley, watch out baby, here comes Spartan."

"Let's have a party," someone of the Goddard crew suggested. "A steak and beer bust." Everyone agreed, and each contributed $8 for expenses. They would meet the Sunday after Thanksgiving at the Wakula Hotel near the Atlantic Ocean in Cocoa Beach, Florida. The Wakula had a propane barbecue grill where they could cook their own steaks.

"If we were only launching this on a rocket, Fred," Windsor said to Wilshusen, "we'd have a lot more fun." Wilshusen grinned.

15

Putting Spartan Aboard

Wilshusen, Stern, and Kohnert were late arriving at Cocoa Beach for the Wakula party. After a delayed takeoff, they finally had left Boulder with its snow-covered mountains, its ski resorts enjoying the first big weekend of the 1985–1986 season. In Cocoa Beach the temperature was 88, everything was green, and there were palm trees along the Beeline from Orlando to Cape Canaveral.

Approaching the Wakula, Stern, Wilshusen, and Kohnert could hear the Goddard field crew already celebrating in the interior courtyard. Just as the Coloradans stepped into the courtyard in the center of the hotel, they were immediately rocked back by a booming explosion, a roaring swoosh of flame spurting 30 feet up, higher than the banana plants around the court. One of the barbecue pits had spasmed. Yells crossed the court. Some engineers dove under the table. Others jumped away, knocking over chairs. Hotel guests, in various states of dress and undress, burst out of their rooms around the balcony.

A couple of men ran to the bar, grabbed bottles of soda, poured them on the grill, and turned off the propane. The flame tongue died hissing. There was general hand clapping and laughing as the party goers emerged from their hiding places like prairie dogs out of their tunnels. The engineers returned to their tables. The danger was over. It was joke time.

"How is it," Stern asked Wilshusen and Kohnert, "that these hot-shot engineers, who have just constructed a high-tech spacecraft, can't operate a propane barbecue?" The engineers retorted they wouldn't trust a scientist to run a propane barbecue either. The happy Goddard hosts tossed cans of beer to the three Coloradans, and they joined the party.

Everybody celebrated. Goddard and LASP had two weeks more of work at the Cape, a thousand little chores that had to be done to prepare Spartan for the shuttle. Unlike the construction at Boulder, unlike the integration at Goddard, they had two weeks' time to do two weeks' work. They had caught up to the flying Comet Halley.

Spartan-Halley was already waiting for them in a giant clean-room, Hanger AE of the Canaveral Air Force Station. These hangers, like everything in Cocoa Beach, go back to the 1960s. Hanger F once held the Mercury rockets, and it's still being used, though now for different purposes. For the Spartan-Halley Mission, NASA set up a few cramped offices with gray steel desks, steel typing chairs, and tangles of phone lines.

The entrance to the clean-room was through the men's room, a heritage from the days when space was solely a man's world. After NASA recruited women astronauts in the late 1970s, women were hired on the lower echelons as well, but the men's room was still used both as a men's room and as the passageway to the clean-room.

Enough protests were raised about that so the security people declared the men's room no longer a men's room. They did not remove any plumbing, but strung a corridor of yellow tape through the place, tape of the kind police use to cordon off crime scenes. Men and women both went along the yellow tape road on their way to the hangar.

Spartan was going "on line," into the *Challenger* cargo bay. As the time got closer and closer when Spartan was to fly into Space, the whole attitude of the crew was transformed. The ritual had changed. It's like the attitude of racers as the race gets closer. They look more critically at their equipment, their skis, their bicycles, their shoes, and even more critically at themselves, their muscles, their breathing, their mind-set and mental readiness. It's not that they haven't been careful all along, but now their awareness is everywhere sharper.

The bunny suits, the shoe coverings, bunny hats, bunny gloves all seemed ultrapristine for the clean-room. The suited up LASP team entered a special entrance room where the door behind them closed magnetically. The floor of this room was a shaker grid, and everyone on it was shaken as if they had suddenly been struck by a nervous disease. As they shook, jets of air shot out of the floor and through their clothes to first loosen any dust particles, then to blow them away.

When the shaking and blowing stopped, the door on the other side was released magnetically, and the team, slightly deranged but purified, walked into the clean-room hangar, 40 feet high, 70 feet long, and 35 feet wide and controlled to a humidity of less than 50 percent. This was high for Boulder where the humidity usually settles into the twenties during the day. Although a super air-conditioning system ran continuously, the LASP people in the hanger continued to perspire.

Several times during the day, quality control personnel carrying flat aluminum plates individually bagged in plastic marched into the hanger and substituted these clean plates for ones that had been distributed around the room on the floor for the past several hours. Later in a lab somewhere, the "dirty" plates were analyzed for anything that might have fallen onto them from the air.

Other quality control personnel on guard duty inside made

sure no one sneezed or dared to take off his or her gloves to grip a fixture more firmly or take off a hat to scratch his or her head. The LASP team never found out what happened if anyone committed such a crime, because no one ever did.

Now the crews began their long series of little jobs. All systems and subsystems were monitored one more time, making certain everything still functioned and that nothing had been shaken loose, disconnected, broken, or gone out of alignment during the careful 365-mile trip from Goddard.

The crews charged the batteries, pressurized the argon tanks that supplied the control jets, and attached and wired the thermal louvers that, in part, controlled the internal temperature of the spacecraft.

Susan Early and her crew from Northrop had measured the Spartan carrier at Goddard for its thermal blanket. Thirteen layers of Kapton and Mylar would further protect Spartan-Halley from the temperature extremes of space. Now the women stretched the covering over the spacecraft, then sprayed it with a special, nonflaking white paint.

It remained only to fasten the decals on the blanket. What decals can be put on a NASA satellite is a matter strictly controlled by Flight Operations at the Johnson Space Center. Spartan could be identified with the "Spartan" rectangle, the "NASA" logo, and the American flag. Nothing else.

Onizuka went quietly to Flight Operations and politely but firmly persuaded the great powers to give him special, nonprecedent-making permission to add a decal of the University of Colorado, a stylized golden buffalo on which the initials "UC" were superimposed.

The powers didn't know it, but Onizuka planned to place his buffalo right on top of Spartan where the Public Information Branch of Johnson would focus their TV cameras during the extensive coverage of the mission. The Colorado golden buffalo would appear on the evening news from KING-TV, Seattle, to Eurovision, Darmstadt.

According to regulations, the flag was supposed to go on the top of the Spartan spacecraft, but Kohnert kept the spot free.

"You can't put the flag there," Kohnert told Early. "That's where the spectrometer radiators are." There was no such equipment, but Early obligingly put the flag on the rear of the spacecraft under the grapple handle.

One night Wilshusen, Kohnert, and Onizuka went into the hangar and affixed the Colorado decal to Spartan themselves. The next morning when the crew came to work, the modernistic buffalo logo was on top of the only American spacecraft designed solely for Comet Halley. Windsor saw it.

"Oh, my God, oh, my God!" He exploded. "This is going to be on national TV flying over the Pacific. Who did this?"

Standing next to Windsor, Onizuka looked like an embarrassed school boy, a little shamefaced, but grinning. His look was a put-on. This was the Onizuka look. By no means was he ashamed about what he had done.

Putting the CU buffalo logo on Spartan was Onizuka's quiet way of putting Colorado on the map in NASA's predominantly MIT, Cal Tech, and Stanford environment. But there was more to the decal than just that. In the spirit of "the little spacecraft that could," Onizuka, Wilshusen, and Kohnert all believed it was a great accomplishment for the University of Colorado to put together the instrument package for the only American Halley satellite.

"How can we get that damn thing off?" Windsor asked. There was no way. In addition to being held with a double stick tape, the buffalo was fixed on the blanket studs holding the blanket. If the decal were pulled off, the blanket would certainly flake, and the spectrometers and cameras might sight on specks in space rather than Comet Halley, and the star tracker might think a speck was Canopus.

Tests, blanketing, decals, inspections, and the chores complete, Spartan now had to be attached to the structure that would support it in the cargo bay, the Mission Peculiar Equipment Support Structure (MPESS).

Looking like one section of a cantilever bridge, the MPESS consists of aluminum girders, a V-shaped mounting which supports a rectangular bridge span. In the middle of this bridge, Spartan is fastened by the two sections of the Release Mechanism (REM). The lower, base section of the REM is permanently bolted to the MPESS, the upper section is bolted to Spartan; the two sections would be latched tightly during launch and then later unlatched for deployment.

On the MPESS, now weighing over 6000 pounds, Spartan was carefully loaded onto a Payload Environmental Transportation System truck (PETS), a job that took over six hours on Sunday. The PETS truck then took Spartan from Cape Canaveral across the Banana River to Merritt Island and to a hanger of the Orbiter Processing Facility (OPF) at the Kennedy Space Center. These hangers are small buildings, dwarfed next to the 527-foot Vehicle Assembly Building which is always featured in pictures of Kennedy Space Center.

Inside the OPF, Stern and Kohnert were surrounded by the tremendous activity and bustle. A loudspeaker continuously squawked information about tests about to take place, hydraulics, latches, pressures. Groups of two or three people, all in bunny suits, and all holding clipboards, walked briskly from somewhere to someplace talking very seriously.

The hustle, the technology, and the organization everywhere around them communicated a charged excitement. It was as if shuttles were landing and taking off in a never-ending sequence. Yet at the same time, disciplined formations of technicians were involved in every level of activity, checking, guiding, controlling. No frenzy anywhere, no tumult but a professional execution of various complicated tasks.

Dominating the hanger in front of the two LASP visitors was a giant maze of structures, cranes, platforms, work stands. They could see only the tail of *Challenger*, rising up out of this metal forest. Stern knelt down on the floor to look at the black underpinning of the shuttle. The orbiter didn't rest on its landing gear but was jacked a few inches off the ground.

"My God," Stern said, "this is a real shuttle!" He had been in the orbiter simulator at Johnson Space Center, but this vehicle directly before him, the *Challenger*, had actually roared into space eight times, zoomed into the upper atmosphere 200 miles up, 80 tons of it, at 10 times the speed of sound, in 2,700 degrees Fahrenheit, then sailed weightless around the planet. To him the *Challenger* was space flight incarnate. Stern was as excited as he had been 20 months before watching UVA 84 blast into the sky above Poker Flat.

Inside the scaffolding, close up, Stern and Kohnert stared at the tiles of the orbiter. The tiles all differed from one another. Each tile was numbered, the numbers clear on the newer tiles, faded on the ones that had gone through several flights. In the rear of the shuttle they inspected the three giant engines. They showed the marks of their use: burnt streak marks from hot spots, the thick snarl of hardware that had to be leak-proof at 420 degrees Fahrenheit and below zero for the liquid hydrogen yet capable of putting out 10,000 degrees of fire and thrust.

Eventually, the PETS truck carrying their Spartan spacecraft backed into the hanger, and after the sides of the truck were removed, a crane lifted the steel cables attached to the four corners of the MPESS, and hoisted Spartan a few inches off the truck bed.

Suddenly everything stopped. A quality control inspector, looked at his clipboard and shouted.

"This crane hasn't been certified in the past 30 days. You can't use this crane." No argument. Spartan was lowered, and the PETS crew raised the sides of their truck again enclosing Spartan in gray metal, all part of one of the 2000 safety checks made before launch.

Stern walked over to the inspector.

"How long is this going to take?" he asked.

"We gotta find another crane," the inspector explained,

"And a certified operator. And do all the paperwork on both cranes. About 4 hours, I guess."

Stern wasn't about to sit around for four hours. He and Watzin went over to Hangar 1 and watched the unloading of the shuttle *Atlantis*, returned from Edwards Air Force Base after its week-long mission, 61-B. As they left the hangar 4 hours later, they saw *Columbia* in the distance with its external tank and boosters, waiting for a launch at the end of December or the beginning of January.

After resuiting, they found Barth and Kohnert back in the *Challenger* hangar, looking up at Spartan and the MPESS hanging from the new certified crane over the open *Challenger* bay. Technicians were in the bay, their bunny suits as crisp and clean as Spartan itself.

One of the technicians in the bay signaled the crane operator on a communication jack. Spartan and the MPESS bridge span moved down quickly until they were about six feet above the men in the bay. The MPESS spanned the bay from left to right with only a small clearance from the sides of the cargo wall.

Now the descent was slow, almost imperceptible, the cargo moving by inches as the technicians checked very carefully, calling for the crane operator to make tiny adjustments. Watching it being lowered, Barth and the others couldn't tell when the downward motion finally stopped, so deliberate was the movement.

The technicians in the bay knew. The MPESS settled in and technicians torqued it into the cargo bay with a complicated trunnion and bolt system. An inspector scrupulously checked each of the fittings, and the MPESS was as solid a part of the *Challenger* as if it had been built into the orbiter at Rockwell.

There was only one more job to do, connecting the electric circuits of Spartan to those of the orbiter and then testing them. Then the bay doors were closed, Spartan shut in, and the orbiter was ready to be transported to the gigantic Vehicle Assembly Building.

Here *Challenger* would be mated to its launching system, the external liquid hydrogen-oxygen tank (though there was no hydrogen and oxygen yet) and the solid fuel boosters. Pointed upward toward space, *Challenger*, securely tethered, then would ride three miles to Launch Pad 39B on the mobile Launcher platform, which is a huge flat truck about 100 feet square moving at 1 mile per hour on four sets of giant Caterpillar tractor treads. The shuttles' majestic crawl has been seen many times on TV.

At Pad 39B TDRS, the rest of the 51-L cargo would join Spartan. The huge tracking and data relay satellite would occupy two-thirds of the bay. Once launched, it would be an important station in NASA's communication network in space. As a matter of fact, NASA higher echelons seemed to show more interest in the launch of TDRS than they did in the launch of Spartan Halley. That made sense. After all TDRS cost half a billion, Spartan maybe a little more than a million.

Before *Challenger*, TDRS, and Spartan could be launched, they had to wait for the launch of *Columbia* 61-C, still on pad 39A where Stern and Watzin had seen it earlier.

16

The Crafty Computer Flight

On January 13, Scobee led five of the 51-L working crew up the steps and into the middeck of the shuttle simulator at the Johnson Space Flight Center. Some 900 miles away, both Spartan-Halley and TDRS were in the actual *Challenger* bay on pad 39B at Kennedy.

At Johnson, the shuttle crew would have their last chance before the flight to practice deploying Spartan-Halley into space and retrieving it. No need to rehearse the launch; the crew had already gone through that several times. There had been lots of all kinds of practices, lots of simulations, but this was *the* "sim," the "integrated simulation," to be conducted with some of the LASP and Goddard crew members under Morgan Windsor's responsibility.

The middeck of the shuttle is not much larger than a walk-in closet. For publicity shots during flight the TV camera is aimed at the ceiling, with the astronauts floating about, weightless. This angle makes the place seem much bigger than it is.

During the launch, the middeck would be even more crowded, with three seats for McAuliffe, Jarvis, and McNair. Christa McAuliffe, "the citizen-teacher in space," and Greg Jarvis had special tasks to perform for the integrated sim, so they were elsewhere. Now the middeck was free of chairs, with not even one set up for McNair.

The shuttle crew filed through the middeck and climbed the ladder into the flight deck. Scobee, as commander, wedged himself into the cockpit seat on the left. Smith, the pilot, squeezed into the one on the right.

Scobee and Smith faced their instrument panel which stretched from one side of the cockpit to the other. It contained all kinds of warning lights, monitors, and myriad instruments you don't find on your everyday Cessna—a Mach speed indicator, for example. Switches were everywhere, on the floor, on the sides, on the ceiling as if Smith and Scobee were in a cave surrounded by chrome lines of tiny stalactites and stalagmites.

The instrument panel is dominated by three computer screens, although the keyboard is on the floor between the two astronauts. There are altogether five microprocessors on board, four of which simultaneously handle every aspect of the mission—prelaunch operations, maneuvering in flight, payload, reentry. Each of the four work together, checking each other; the fifth is a backup for the others, redundancy again.

During launch, in the area behind Scobee and Smith, there would be chairs for Resnik and Onizuka, but right now Resnik and Onizuka were standing with McNair. On the back wall toward the cargo bay is another swath of toggle switches, all stiff enough so that a astronaut can't activate one simply by brushing against it. At the base of the switch panel are two small joy sticks and a small notched control wheel that Resnik would use to operate the robot cargo arm.

Don't think the simulations aren't real. They are. This last rehearsal would be as close to the actual flight as 25 years worth of evolution in simulation technique, updated from the lessons of 24 shuttle flights, could make it. The simulator duplicates the shuttle in every detail. The crew experiences all the sights, sounds, even feelings (except weightlessness) of actual flight. The windows of the simulator are, in reality, high resolution television screens with almost twice as many lines as home TV receivers and so transmitting with twice the clarity.

During sims the astronauts can see the Earth as it would appear from orbit—wisps, stipples, striates, and commas of clouds. The sky shows a thousand stars all in proper magnitude, the actual constellations in their actual positions, so Smith could use them for navigation. On the Earth beneath, swatches of green amid browns, golds, and the crumbled, lined gray of mountains. It is all so accurately projected that the astronauts could pick out wherever they were: Hawaii, Florida, Italy.

Operating the simulator is a roomful of computers and digital image generators that respond to input from the astronauts exactly as if they were on orbit.

The realism of the simulator is absolutely convincing. Astronauts know it's a simulator, sure. After all, they're not weightless, but the only reality astronauts experience is the interior of the simulator-shuttle with the Earth and black sky outside provided by those TV windows. The effect is hypnotic.

Even in a darkened movie theater you forget where you are. You're not in a theater, you're in Rick's place, Gutman's apartment, or on the River Kwai. When the lights come on again, you're jolted back into reality.

The imprisoning and enveloping effect of the simulator is several degrees more convincing than that of the darkened theater. Astronauts aren't jolted out of the shuttle world until they leave the simulator and see the bleak interior of the Space Simulation Laboratory.

Everything on the simulator (as it is on a real shuttle orbiter) is connected to the Mission Control Center by telemetry—all the instruments, monitors, control systems, computers. Mission Control is across the space center from the simulator, but in order to achieve a realistic time lapse, signals go from the simulator to a satellite, then down again to Mission Control.

Mission Control for the sim was just as it was for Skylab, for the previous 24 actual shuttle flights, and for the upcoming *Challenger* flight 51-L—the two-dozen member flight control team at CRT's in the theater, with the 200-person support team unseen but communicating with them from rooms surrounding the theater.

On the front row sitting in front of the huge screens are the flight dynamics officer (FIDO), guidance and navigation officer, guidance and control officer, and flight trajectory officer. These all guide the flight path and trajectory of the shuttle on the basis of the information sent from both the shuttle in space or the simulator.

The second row watches out for shuttle systems: here payloads officers are backed up by the customer support teams with their own computer in a room off the theater. Beside them are environmental control, life support systems, communications systems, data processing. Behind these in the third row is the capsule communicator (CAPCOM), an astronaut and the only member of the flight control team who is in voice contact with members of the crew flying aloft.

The flight director, Randy Stone, captain of the team, stands next to CAPCOM, and through CAPCOM he channels whatever has to be said to the shuttle crew. More people are on the floor of the operations room, but they are primarily advisory.

Every minute that the sim is in communication with Mission Control, a million bits of information reach the support rooms. Here the backup specialists evaluate, synthesize, and relay the information via computer to the appropriate officer in Mission Control. The officers reevaluate the information they receive and, when necessary, communicate with the flight commander through the CAPCOM.

Stern and Kohnert flew from Boulder and joined Windsor, Watzin, and more of the Goddard team in the customer support room. Seated around an elongated, U-shaped table, they listened over the intercom to the voices of the flight control team in the theater. They could hear the quiet voice of the flight director in the theater, but they couldn't reply by voice. In fact, they could reply only by sending a written message shot by pneumatic tube, the P tube, they called it, to a specific officer on the floor.

Behind all the simulations is the crafty brain of the simulation supervisor who for each sim acts in place of "The divinity that shapes our ends." The supervisor writes all kinds of problems into the scenario of the mission—little problems, big problems, nuisance problems, even technical problems that test Ph.D.s in electrical engineering. He creates special situations that reproduce the chance equipment failures, false instrument readings, and technical malfunctions that cause delays or that without accurate diagnosis and correction might even cause the mission to fail. The supervisor inputs the problems into the simulator instruments and computers. As in a normal flight, these malfunctions are also signaled to the consoles of the flight controllers and to that of the support staff.

The simulation supervisor is not just trying to keep alive the myths of demons and gremlins. Since the sim is a training technique, it has to be tough. The astronauts and flight controllers have survived diabolical sims before. Each orbiter, however, has different peculiarities; each payload has different requirements and demands different skills. The astronauts and flight controllers go into the sims ready for shuttle flight. They emerge ready for any problems the payload may offer. Since the supervisor is testing the crew's and flight controllers' handling of the payload, at the same time he also tests those in the customer support room who, like LASP and the Goddard teams, provided the payload.

The sim for Spartan-Halley began after *Challenger* was theoretically two days into its flight. Scobee, following a multipage checklist, flicked over a hundred of the stalactite and stalagmite switches, and they'd better be just the right ones. He punched columns of figures which registered on all five computers at once, then checked the gauges and instruments across the panel.

Down in mission control, FIDO figured the shuttle's exact position and CAPCOM sent it up. Figures rolled on the computer consoles, including the CRT at the head of the table in the customer support room. The sim was operating.

On orbit the shuttle was "upside down," and through the windows crew members saw the Earth over their heads and the stars beneath their feet. But who feels it? In space they're weightless, and as on Skylab there's no up or down.

Spartan was snug in the cargo bay, inactive. Resnik, Onizuka, and McNair looked out the two windows toward the cargo bay. On order from CAPCOM Resnik opened the bay. They saw Spartan beneath them, the CU decal on top, the grapple handle to their left.

Resnik was supposed to move the robot arm so the TV camera on it would allow Onizuka to inspect Spartan in great detail to make sure the spacecraft had not been damaged in launch and early flight. Earlier he and Stern had gone over the satellite and listed all minor scratches and nicks so Onizuka would not send unnecessary information to Mission Control.

Now the supervisor struck with one of his scenarios. A circuit in the arm elbow was malfunctioning, and Resnik had to try several circuit bypasses until she got it functioning correctly. Finally she moved the camera in place. Onizuka saw no frost on the edges of the sunshade, no dust particles, no chalking of blanket paint, no new scratches or nicks.

Actually, of course, Onizuka was watching a video of Spartan, shots for the sim supervisor. The supervisor could have painted in some damage as part of the test, but he didn't.

McNair now punched out his hour-long check list for the second time.

"No problems," said Watzin smiling.

The next activity would occur at Mission elapsed time 24 hours, and the Goddard-LASP customers had a 30-minute break that represented the passing of that time.

To the customer support room, FIDO sent the exact position of the orbiter for deployment of Spartan so all the highs and lows could be worked out and teletyped to McNair. McNair and his teammates had all done this before at Goddard, and the operation went smoothly.

Resnik could now deploy. She could have programmed the arm ahead and deployed by computer, but she wanted the experience of deploying manually in case an emergency arose on the actual flight.

The robot arm responded with correct movements and even with sound as Resnik flipped the switches and operated the controls manipulating the shoulder, elbow, arm, wrist, and grip.

It was noon in Spartan's orbit. Resnik placed Spartan gently in space with the Sun sensors facing the Sun and the star tracker facing generally toward Canopus. She released the carrier from the arm. Scobee reported that Spartan pirouetted. Smith flew the orbiter away from the spacecraft, and the Spartan-Halley mission was underway. No need to repeat the 43-hour test which had taken place at Goddard. The teams had confidence in the little spacecraft that could.

After a short break, Stone gave the order for retrieval. The simulation supervisor struck again. The shuttle was supposed to locate Spartan by radar, but the radar didn't function correctly. Scobee maneuvered the shuttle until, with its acquisition lights, it picked up reflectors like bicycle reflectors that had been fixed on Spartan for just such an emergency.

Carefully watching velocity and direction, Scobee moved next to the Spartan, his 90 tons of orbiter close enough to the 1-ton box that Resnik could pick it up. This was something like a 12-wheel truck and trailer barreling down Donner Pass and the driver asking the passenger to pick up a basketball rolling beside them. And Resnik had to be as accurate as a brain surgeon. If she nudged Spartan at all, it would move away.

Everything worked. She grappled with the box, lifted it up and hung it over the bay. But the temperature on the release mechanism (REM) registered minus 100 and falling. Resnik could not lower Spartan because the REM motors wouldn't operate at that temperature. Spartan was just hanging there, but it could not hang there much longer. Stone sent Windsor a choice.

"We can't latch. Do we abort?

Windsor addressed the group around him.

"You mean they'd dump Spartan just for this? What happened to EVA? I got one of those damn booklets on extravehicular activity. According to that, McNair and Onizuka can go into the bay and lower it themselves."

Windsor turned to the secretary who wrote directions for the P tube.

"Tell 'em EVA," he ordered.

On the computer consule in front of him, Windsor and the rest saw the temperature drop to minus 120. A message came back.

"Negative EVA. Too cold. Do we abort?"

"This is unreasonable." Windsor said. "On other flights the coldest equipment ever got was minus 60 or 70. The latches should have been tested for lower temperatures. They can't just dump our bird. We have some beautiful stuff on board."

A representative there from Mission Control to advise the crew explained. Sure this was a sim. But test or no test, the latches might very well fail in flight. On the shuttle there's a tremendous temperature range from fore to aft.

"The carrier has to be latched to the MPESS," he continued. "You can't have a one-ton box jumping around the cargo bay. NASA's not going to risk wrecking a shuttle or TDRS for your box, Morgan, beautiful or not."

Someone at the end of the table interrupted.

"Tell them barbecue mode, Morgan."

"Barbecue mode! What in hell is that?"

"The CDR, Scobee, rolls the orbiter so the cargo bay is in the Sun until it warms up enough. You know, like barbecuing a ham on a spit. You rotate it."

Windsor shook his head incredulously, dictated "barbecue" on a slip of paper, and the secretary shot it through the tube.

The orbiter rolled. In 20 minutes the figures on the console started to rise. Everybody cheered. That was the solution the supervisor had programmed.

Signals finally came down to the computers that Spartan was berthed and latched, then binary O's and 1's moved across the consoles. McNair, on his APC, was asking Spartan "How did it go?" All during the flight the Scotty computer kept track of what went right and what went wrong inscribing it on the tape-recorder every few seconds.

"Were you powered?"

"Yes."

"Were you deployed?"

"Yes."

"How many times did you scan?"

"Twenty-three."

It was a long mission-success-word program, and as they translated the binary numbers that were on the screen, everybody knew the mission had gone reasonably well. Without that program no one could know for sure for 10 days, until the end of the entire *Challenger* mission plus unloading if something had gone haywire.

In addition, Mission Control had to know quickly whether Spartan had failed, for LASP had saved 8 hours of tape for a backup, the in-the-bay mode. With Spartan still in the cargo bay, Scobee could maneuver the orbiter and point the spectrometers toward Comet Halley and accomplish at least 8 hours of the mission. Yet the procedure was dangerous, a desperate alternative. Stone and Scobee both were glad the diabolical simulation supervisor had not required the in-the-bay mode.

The sim was over. The three teams, the astronauts in the simulator, the flight control team in the operations theater, and LASP-Goddard personnel in the customer control room, were given a 10-minute break, then all of them were brought together on the communication channels for a debriefing. Everyone had a chance to say, "Here's where I had a problem." "Here's what I learned." "We could do better if we did this."

Scobee was critical of the H and L relay system for updating the Spartan computer.

"It's a waste of valuable time," he said. He suggested that the Spartan carrier, regardless of what experiments it carried, should be a full-fledged satellite with information telemetered to and from the ground.

"Then you could learn what it was doing, how it was going without tying up a mission specialist during a flight." Scobee and Smith both promised to be advocates with the hierarchy for changing the Spartan system.

The final, integrated sim is traditionally followed by a party. To celebrate, Stern and others went to the Merida off of I-45, the Gulf Freeway, for margaritas and nachos and from there to Appleby's for beer and popcorn with McNair and Onizuka.

Everybody toasted Spartan asleep in the *Challenger* bay, and usually with the same toast, "Here's lookin' at you, kid." The LASP-Goddard engineers and scientists toasted the astronauts and controllers, "Louis, I think this is the beginning of a beautiful friendship." On Challenger flight 41-B, in 1984, McNair played his alto sax. On 51-L he promised to substitute for Dooley Wilson's piano and play, "As Time Goes By."

This *Casablanca* binge and the bad imitations of Humphrey Bogart came from the fact that Casablanca, Morocco, had been designated as the alternate to the alternate landing site. In case of emergency and if Scobee couldn't bring *Challenger* in on either the Edwards or Kennedy landing strips, he was to glide 4000 miles across the Atlantic to Dakar, Senegal. However, the weather was already getting bad there, and Casablanca was its alternate.

The launch of *Challenger* had been set back from January 22 to the 24th. Early on Tuesday, January 21, Stern was taking a shower and preparing to leave for Kennedy. The phone rang. He ran from the bathroom to his office, dripping and shivering. It was McNair on the phone. He told Stern that as a result of the integrated sim, he had written down 15 or 20 changes in the deployment and retrieval checklist that would make both procedures run smoother and faster.

It's a very complicated process to change the checklist of a shuttle flight. Changes involve the flight crew, Mission Control, and the experiment scientists. Weeks before McNair and Stern had spoken to Scobee about changes.

"Forget it," Scobee said. "If it works the way it is, we're not going to make it better. If it ain't broke, don't fix it."

"Have you spoken to Dick again?" Stern asked.

"No," McNair answered. "It's just between us. We got to work this out ourselves. I've got my own checklist. I'm going to carry it in my hand during the flight. You and I are going to get these things straight, and it's going to be our little secret." They went over McNair's list for two hours, making adjustments as they went along, Stern writing down the changes they agreed on. In case anything broke down while McNair was punching in the information, he would ask the CAPCOM to put Stern on a special communication line.

After McNair hung up, Stern was struck by the incongruity of what happened. He was naked, wet, and cold, talking to an astronaut who, in a few days, would be flying around in space at 18,000 miles per hour. Unknown to the flight commander, they had worked out a secret checklist.

When he was in high school dreaming about being an astronaut, Stern had envisioned it as all glamour, never something like this. As he dried himself off, Stern wondered if Alan Shepard or John Glenn changed flight plans at the last minute. Scobee and Stone would certainly notice the changes. McNair would be put into some kind of NASA doghouse. And so would Stern.

17

The Last Flight of the *Challenger*

Later that same day, January 21, 1986, Stern and Kohnert flew to Kennedy for Red Tag Day. Every object that was unnecessary to the experiment or could come loose during launch or flight had to be removed from Spartan. Each of those objects had a large red tag hanging from it.

The removal of these tagged items was very carefully checked. Someone from Lockheed has a data bank of the red tags on all the satellites launched on shuttles. He appears a few days before every launch, and his only job is to go around with his list inspecting to make sure nothing tagged is left in the shuttle bay.

Stern and Kohnert met the Lockheed man on Pad 39B at the base of the "Pig 'em," payload ground handling mechanism (PGHM), a 130-foot movable tower. This structure can swing up next to the orbiter, where doors at the sixth story level can air-seal to the bay doors, forming a continuous clean-room.

Before Stern and Kohnert could enter the PGHM elevator, however, everything they were taking into the bay had to be fixed to their bodies. NASA has a very long, exact checklist that has to be followed in every detail. Your glasses are strapped to your head. Tape goes over and around your wedding ring and watch.

Stern and Kohnert actually strapped dog collars on each wrist, and whatever they carried, a pencil flashlight, a squeeze bulb, or a ballpoint, whatever, was wire-looped to the dog collar. NASA has developed a special procedure for Q-tips. Stern carried some in a tethered plastic bag, but he had to have the same number in the bag when he left the cargo bay as when he went into it. In transferring anything from one wrist to another, they had to loop the object on the right wrist, before it could be loosened from the left.

Dropping anything would not endear you to NASA. Whatever you dropped had to be found before launch, even if it meant moving the shuttle back into the hangar and removing the payload from the cargo bay.

During the week that *Challenger* was on the pad awaiting launch, a storm of hurricane intensity had rocked the shuttle on the gantry "like a palm tree," the newspaper, *Florida Today*, reported. The winds had broken the seal of the cargo bay, and while a protective cover had kept the instruments of Spartan from damage, Kohnert noticed waterspots on the sunshade, and carefully removed them. Then Kohnert checked the current from the instruments to the shuttle cockpit, which flowed out through a huge rope-like harness. McNair made sure the ion pumps for the spectrometers were working, because to maintain the vacuum they had to operate twice a day all the time the shuttle was waiting to launch.

At the same time Stern checked his own list, making certain nothing red-tagged was left, and the Lockheed man checked him. Then the cargo bay could be resealed. The LASP Team, Wilshusen, Stern, and Kohnert, planned to watch the launch and afterward fly over to Johnson Space Center and join Neil, White, Windsor, and the rest in the customer support room of Mission Control.

Barth arrived at Kennedy with the new president of the University of Colorado, Gordon Gee, and a group of committee chairpeople from the Colorado legislature, all eager to see the launch of Colorado's first shuttle payload. Gee wanted to show his support of CU's astronaut alumnus, Colonel Onizuka, especially after Onizuka's successful ploy with the CU buffalo decal. At the same time, Gee and Barth both stressed to the politicians and to the chairperson of the state's all powerful joint budget committee how closely CU was involved with the prestigious shuttle program.

David Aguilar from Ball Brothers brought a group of Colorado school children with some of their teachers to the launch for McAuliffe's "Teacher in Space" program.

Challenger was supposed to launch January 24, but a sandstorm at the alternate landing site, Dakar, caused a delay. The TDRS people figured that, in case of a strong headwind in Africa, the orbiter could not glide the extra 750 miles to Casablanca. Remove Spartan and MPESS, they advised NASA, and the orbiter would make it.

When Barth heard this, he flew as fast as possible from Kennedy to Houston and finally convinced flight operations management it would take four days to remove Spartan, and on the authority of the National Center for Atmospheric Research at Boulder, Barth assured operations the dust storm would end at Dakar in close to 24 hours. Spartan stayed.

Saturday the 25th, rain and cold. Gee and the legislators surrendered to the weather and returned to Colorado. The schoolchildren, on the contrary, braving the rain and cold, remained at Kennedy. They learned that although commercial airliners can take off in the rain, the shuttle can't, because rain can erode the tiles. Actually, the shuttle is not allowed to take off if rain clouds are within 30 miles of the pad, or if there is less than a 7-mile visibility. Vice President George Bush arrived, leading a parade of overcoated administrators who inquired about the reasons for the delay. These were explained. On Superbowl Sunday, the schoolchildren, after a simulated countdown, settled in front of the hotel TV sets.

Monday the 27th was clear, but cold and windy. The children could see *Challenger* on the pad 4 miles away across the Banana River. Waiting for 4 hours, they huddled on a small hillock, some wrapped in hotel blankets and others in green plastic garbage bags, which provided no warmth but at least cut the wind.

They didn't know that computer consoles in the launch control center had signaled that the latch to the middeck hadn't locked properly. It was stuck. It could neither be closed further nor opened at all. It was fixed to the fuselage by titantium bolts which operators ordered drilled out. The titantium bolts chewed up drill bits and hacksaw blades before they were finally removed. Then the wind increased to 30 miles per hour. Launch was delayed until Tuesday, the 28th.

On Tuesday the children again waited 4¾ hours, the luckier ones in blankets, the others still cold in plastic bags. Kohnert and Wilshusen were in the crowd near them. Barth and Stern, wives, and other special guests waited in bleacher seats a mile closer to the pad. To keep warm they kept going to the cars or buses near the bleachers where heaters were running. Then the countdown reached T-10 and counting. The shuttle was a go.

Finally, the children saw the great rolling flame of the launch. The brown pelicans and other birds of the Merritt Island Wildlife Refuge heard the rumbling, thunderous sound before it crossed the river, and they scattered through the air between the flame and the enthralled children.

As the children watched the roaring clean white line rise from the flame, it suddenly burst into huge balls of flame and great white clouds.

"Oh my God," said one of the children.

"That's it?" another asked. "Is that what's supposed to happen?"

"That's supposed to happen, isn't it, Mr. Dennler?" one of the sixth graders asked his teacher, Bill Dennler.

"They're the boosters," another student assured him. "They'll come down. You watch."

In the Mission Control room at the Johnson Space Center, the numbers transmitted to the two huge screens on the left- and right-hand sides of the world map suddenly blinked. Rows of X's displaced the numbers. The communications officer thought he had merely lost the signal. Possibly one of the ground antennae had failed. In the communications support room the technicians tried to switch to another station and reestablish contact.

On the Mission Control floor the flight director told the communications officer, "They're in the next step," meaning the crew of *Challenger* had followed the standard procedure for loss of communications. They had rehearsed it many times.

"Yeah," the communications officer answered. "They're on cue cards," meaning the crew was flying on its own.

In Mission Control, there is no commercial TV following the launch and flight. There was a TV in the Spartan customer support room. White, Windsor, and the others saw their computer screen go blank. On the TV screen they saw the ugly swollen cloud.

White couldn't tell from the picture what the cloud was. It had actually been caused by the explosion of the two solid rocket boosters. He kept hoping the orbiter would come through the cloud of smoke. He was waiting to hear Scobee say, "Houston, this is *Challenger*, such-and-such happened, but we're OK."

Windsor with his rocket experience knew what had occured as soon as the CRT went blank. A rocket only loses contact with the ground when it has been destroyed. He didn't have to watch the TV. Immobile, he stared at the empty computer screen.

After an eternity of 1 or 2 minutes, on the floor of Mission Control the flight dynamics officer stood up at his console.

"We have lost our family," he said.

The flight dynamics officer knew that the seven astronauts had to have perished from the sheer force of the aerodynamics, the terrible air pressure that had been exerted on the orbiter flying at 2000 miles an hour, 3000 feet per second, then being suddenly thrust into a very sharp curve.

That air pressure tore off the doors of the cargo bay and ripped the Spartan spacecraft out of the shuttle bay. Spartan was torn off the release mechanism, 32 two-inch hardened steel bolts were stripped out of the bottom of the carrier.

The shuttle itself, in pieces, twisting and spinning, fell back to Earth bearing its destroyed heroic crew. Colliding with the surface of the Atlantic off the Florida coast, the shattered debris sank through the saltwater to the murky depths.

Spartan had been thrown free of the shuttle. The white Kapton blankets wrapped around Spartan were snatched and propelled into the turbulence. Spartan continued to hurtle toward Earth. When it hit the thicker air of the atmosphere, the increased friction turned the front of the carrier box into a glowing meteor. The heat transmuted the metal into the consistency of plastic. The rear, apparently in shadow, did not get as hot nor did it change its metallic characteristics.

The carrier, still intact, hit the water of the Atlantic at 400 or 500 miles per hour. At this velocity the water might as well have been a rock.

Several months after the crash, in the spring of 1986, all that remained of the *Challenger* and its heroic crew had been retrieved and returned to the Kennedy Space Center under the tightest possible security and put under close guard in one of the hangars. A grieving nation continued to mourn.

Spartan had been raised from 60 feet deep in the ocean and carefully inspected. Spartan's sunshade sheared off. Only a third of the original contents of the carrier remained within the mutilated box. What had been meticulous precision within the instruments had buckled into gross chaos. One spectrometer was gone from the carrier. The case and deformed shell of the other was shoved back on bolts bent like wet bread sticks. The spectrometer's telescope had been torn off. A camera had exploded out of its box, its film still rolled as if ready to photograph the stars and the comet. The heavy batteries, which were to provide life to the instruments, remained. The star tracker, like an ancient castle tower, deteriorated not in centuries but in an instant. Of the microprocessors only a small piece of a circuit board was left.

Water had seeped between the lenses of the star tracker, behind what were once tight seals. With the water came thousands of barnacles that fastened themselves to the metal. Some barnacles, like pieces of thick thread, all the color of pus, crawled into the shattered spectrometer.

Immediately after the explosion, at the Kennedy Space Center, loudspeakers could be heard all over the launch site.

"Ladies and gentlemen, may I have your attention. We have had a major malfunction. We ask that you quietly calm the site" (the announcer's nervousness caused him to falter) . . . "quietly clear the site and go calmly to your cars."

Some reporters had seen the children and flew at them like a cloud of insects buzzing with questions. Dennler gathered the sixth graders and guided them away toward the bus, answering their questions as honestly as he could. They all knew what had happened but did not want to accept it.

This was Kohnert's first shuttle launch, but he had seen five rocket launches. He thought the shuttle rocket had jettisoned an earlier stage like the two-stage Taurus-Orion at Poker Flat, but jettisoned too soon.

"We should still be under rocket power," he said to himself. "We're not going to get our mission right." Kohnert, like the schoolchildren, wanted to deny what he knew. He was an aerobatic pilot who knew the limits of aircraft. As he observed the explosion, he knew the orbiter had been forced beyond the limits of stress by the aerodynamics of the atmosphere.

Wilshusen, like Windsor over at Houston, realized immediately what had happened but was too shocked to say anything. Kohnert and Wilshusen drove in silence to their motel down A1A, imprisoned for 3 hours in heavy traffic.

As soon as they returned to their rooms, Kohnert turned on the TV. As he listened, he thought that during the long drive he had missed what happened to the shuttle, that it had landed some place. He didn't know the networks were broadcasting the launch over and over again. He heard what the astronauts were saying and what Houston was saying. Why were they saying that? Where was the shuttle?

Wilshusen was still silent. For a time he had been thinking of retiring from LASP. Now he knew he would.

Above the Cape a white contrail halted, twisted, then fell in gossamer ribbons. Streamers of white vapor descended, as if death had continued on the path of its own inertia. The streamers were slowly torn up in the wind, twisted and mangled by the air currents. As Barth left the stands with his wife, reporters shot questions at him, but Barth never expressed his feelings in public.

"I don't want to talk about it," he said. He told Stern and Carole he wouldn't go back to Boulder right away. He wished to get away to a place where he could be alone.

A reporter from Denver drove Carole and Alan Stern back to their hotel. They stood by A1A and watched his car. It seemed as if he was going toward the Beeline back to Orlando. Suddenly he just looped around and headed toward Kennedy as if he had realized, "My God, I'm a reporter. I have a story to cover."

Stern and Carole Stern flew to Houston. It was home for both of them. Carole had been raised in Houston, and Stern had felt that going to the space center was going home too. He had remembered again the "I'll be back" scratched on the underside of the Saturn V. He knew the message was no longer visible. The maintenance department of Johnson repainted the rocket each year to keep it in mint condition.

As he met them one by one at the Johnson Space Center, Stern thanked the members of the flight control team for their work with Spartan and an earlier experiment of Stern's carried on *Columbia* 61-C.

They all talked of 60 seconds. If the fuel leak had burst 60 seconds later, the shuttle would have been at 200,000 feet, in the near vacuum of space. There would have been no tumult of air, no such terrible violence. The astronauts would have survived.

Neighbors, friends, fellow workers of the astronauts from the surrounding areas, from Clear Lake, Baybrook, Seabrook, Ellington, Houston, from everywhere, brought flower wreaths and bouquets and laid them on the lawn of Johnson Space Center, many next to the Saturn V or to the two rockets nearby, the replica of Alan Shepard's Mercury-Redstone and the test rocket for the Apollo. This first, spontaneous outpouring of grief for Francis Scobee, Michael Smith, Judith Resnik, Ellison Onizuka, Ronald McNair, Gregory Jarvis, and Christa McAuliffe turned a huge area, larger than a football field, into a garden of remembrance.

Invisible 93 million miles away, Comet Halley followed its inexorable orbit.

18

The Halley Encounters

From those last tragic days of January until the middle of February the Comet was invisible on Earth. In space, Giotto, the Vegas, Suisei, and Sakigake were soaring toward it. Giotto and the Vegas, on modern voyages of discovery, would pierce the mist of the comet's atmosphere to reveal something of the unseen world until now hidden within, hidden from Attila the Hun in 451 A.D., from William the Conqueror in 1066, and from Edmond Halley, the astronomer in 1682.

Giotto and the Vegas would measure the size and shape of the comet's flying nucleus, the nature of that surface, and the nucleus' rotation (for all objects in space revolve). The other exploring craft, Suisei and Sakigake, would measure the composition and growth of the atmosphere that hides this nucleus world, measuring it no longer through the distorting veil of the Earth's atmosphere but across the near vacuum of space.

During the last week of January and the first of February Comet Halley soaring at more than 100,000 miles per hour was only six days from its closest approach to the Sun. This was the window during which Spartan-Halley should have been watching the comet. It would have been the perfect opportunity. Comet Halley was now providing a real show. After a relatively quiet period, a series of outbursts began, consisting of spectacular jets and explosions. Stern logged on his computer and disheartened and melancholy saw the reports of a comet gone berserk.

The comet was 25 million miles above the planet Venus. The controllers of the Pioneer Venus Orbiter at Mountain View, California, fired its thrusters in 100 half-second pulses, and the spacecraft's ultraviolet spectrometer shifted from the planet and onto Comet Halley. These observations had been planned to complement those of Spartan-Halley. Half of the evidence, at least, would be available.

Pioneer Venus mapped the hydrogen cloud extending 15 million by 21 million miles around the coma, which made Halley, with its atmosphere, the largest object in the solar system. Every day Pioneer Venus transmitted to computers the coma's mounting temperature, the evaporation of water molecules, the velocities and ionization of the atoms hydrogen, oxygen, nitogen, carbon, and more.

At perihelion, the point of the comet's orbit nearest the Sun, the comet was never directly behind the Sun, as it seemed from Earth, but rather 6½ degrees above the Sun, and some 54 million miles away. For the first time ever, Pioneer Venus told astronomers what happened when Halley was scorched by the full blast of the Sun's 10,000 degree Fahrenheit temperature, by the outpouring of the 4,600,000 tons of mass the Sun explodes into space every second. Ian Stewart, in his office at the University of Colorado, calculated that Comet Halley lost between 45 to 54 tons of gas and dust every second it was exposed to the Sun's onslaught.

During all the time the ultraviolet spectrometer was transmitting its data to the laboratory in Boulder, Stewart found what seemed to be a 7-day variation in the outflow of gases. To the scientists at Colorado the variation suggested the comet might be revolving in a 7-day cycle, a comet's day that was an Earth-week long, much longer than the 44 hours Stern had figured for the doomed Spartan-Halley mission. Without the data that Spartan-Halley would have collected, scientists as yet had no corroborating evidence for the comet's period of rotation.

Pioneer Venus followed Comet Halley on its trip until March 6, when the comet was 125 million miles from the satellite. By this time, Stewart had received 20,000 separate scans, the greatest accumulation of data yet to arrive on Earth about the comet.

On the other side of the Sun, the first of the comet explorers, Vega 1, had entered the mist of the coma. When it was still 8 million miles away, Vega was guided toward the comet by its two cameras. The first, the scout camera, was programmed to fix on the brightest spot in the comet's atmosphere. This bright spot, however, was not the nucleus, as the programmers assumed it would be, but the jets of gas and dust exploding out of the nucleus.

Vega transmitted the first pictures of this bright spot, one every 20 seconds, its signals traveling first to a 230-foot-wide radio dish at a station in Crimea, from there to the computers of the IKI in Moscow.

Binary numbers transmitted from Vega were converted by computers in Moscow into color, each color a different intensity of brightness. The huge screens in the viewing room at IKI showed bands of purple, blue, green, and yellow fanning out from a flower-petal center of red.

This image processing technique serves the same purpose as maps on TV weather shows, where blues, yellows, and reds represent intensity of rain. The contrast of colors was dramatic and made the differences in degrees of white, gray, and black easier for the audience to distinguish.

In the instrument operation center of the IKI all principal investigators with an experiment on the Vegas watched the figures coming onto their monitor screen, each computer identified by a sign, its Cyrillic acronym, MISCHA for the magnetometer; IKS, the infrared spectrometer; PHOTON, the shield detector; and DUCMA, John Simpson's "dust particle counter and mass analyzer."

The operation center was completely international. To exchange information verbally, the PIs used English accented by French, Hungarian, German, Russian, Austrian, Chicagoan, and Bostonian. However, the information that they exchanged was exciting enough to overcome even the heaviest accent.

All along the trip from Planet Venus, the instruments on Vega were activated from time to time for testing. DUCMA intermittently reported the interspatial dust, three general sizes of particles, some large, but still smaller than the tiniest grain of sand; some medium, the size of the dust in cigarette smoke, and the smallest a fraction that had to be written as one over 10 plus 16 zeros of a gram. Still particles traveled more than 48 miles per second when they were caught on DUCMA's electric tape.

ICE, six months before, had swooped into Giacobini-Zinner's tail behind the nucleus away from the Sun. In contrast, now Vega 1 entered Comet Halley in a different part of the comet's atmosphere, in front of the nucleus and toward the sun.

Traveling about 162,000 miles per hour, Vega 1 crossed through the bow wave of the comet, an area of turbulent gas, dust, and a magnetic field caused when the solar wind, moving at 864,000 miles per hour, wraps itself around the nucleus and coma.

Bow wave, taken from the action of ships in water, isn't the best analogy here. With ships, of course, the bow wave is formed from the bow of the ships cutting into the water. Comet Halley, however, was now on its flight away from the Sun—outward bound back into the frozen depths beyond Neptune, its aphelion, its point furthest from the Sun. Therefore, the solar wind and radiation were coming from behind, and the comet was a white-water raft sailing downstream or better still a submarine, because the bow wave completely enveloped the comet. The Sun's waves were making a stern wave. Astronomers know all this, of course; nevertheless they still use the term *bow wave* or *bow shock*.

The instruments began relaying their information about 2 hours before the Vega was scheduled to make its closest approach to the nucleus—5,500 miles away. For these close-up shots, the second of Vega's cameras went to work, imaging from different angles as it sped past. The audience watched a red flower petal elongate into an egg and the surrounding bands of yellow, green, and blue take on the oblong shape. Finally the egg turned into something peanut-like, thinning from the bulbous end then enlarging again slightly at the other end.

The number of dust impacts increased to 20, 30, 50 hits a second, and more of them were larger particles. Then the impacts ran up to 1000 a second, and more, beyond what the DUCMA could register. Simpson believed the Vega had gone through a panorama of dust 100 miles wide.

Sagdeyev had been encouraged when ICE had glided through Giacobini-Zinner with a minimum of damage from the dust, but he didn't want to take any chances with Vega. After all, the Vega solar panels weren't wrapped close to the spacecraft as they were on ICE. On Vega they stuck out like airplane wings.

During Vega's charge through the dust and gas of the bow wave, the controllers put the instruments on batteries. It was lucky that they did, for the dust tore up the solar panels, and Vega 1 lost 45 percent of its power input. There was other damage as well. A spectrometer was knocked out and a few instruments were crippled, but the rest functioned well.

All in all, Vega 1 sailed for 20 minutes, or a distance of 60,000 miles, through the comet's atmosphere. Sagdeyev was elated and sent word to Gorbachev, who was then presiding over a meeting of the Central Committee. In three days, on Sunday the 9th, Gorbachev announced, members of the Central Committee would witness the trip of Vega 2 through Halley.

Meanwhile, Suisei and Sakigake were approaching Comet Halley. Suisei, the pioneer, sailed from Kagoshima to cross 180,000 miles behind the comet as it sped away from the Sun. Then on the day before Vega 2 was due to arrive near the comet nucleus, March 8, the controllers in Tokyo fired jets to push Suisei 10,000 miles closer to Halley. At this distance, 170,000 miles away, no one expected that the 308-pound Suisei would be bothered by the dust particles, but in fact the little spacecraft was thumped by two giant particles which rocked the craft a little. Suisei righted itself, however, and sailed on smoothly.

Since November 1985, Suisei had actually been monitoring both the coma of Comet Halley and the hydrogen envelope outside the coma, the data being picked up on NASA's deep space network in Madrid, Spain, and Goldstone, California. Tomizo Itoh, the mission director, was fascinated by pulsations within the coma that seemed to have a 52-hour cycle. The pulsing was strongest when their instrument registered that area of the coma as being hotter, less when it was cooler. The rhythmic pulsations suggested to Itoh and the Japanese scientists that Comet Halley had an active side and a passive side; possibly here was evidence that the comet was rotating every 2½ days and not every seven days, as the evidence from Pioneer Venus had seemed to indicate. The possibility of a 2½-day rotation was closer to what Stern had estimated for the Spartan-Halley mission.

On March 9, while Suisei narrowed its approach to Comet Halley, in Moscow members of the Central Committee of the Communist Party and of the Presidium arrived at IKI in their Zil limousines. They were immediately guided to a special room off the main viewing room where they had a clear view of the video screens as they waited to follow Vega 2's trip through Halley's atmosphere.

Sagdeyev had declared March 9 "Christa McAuliffe Day," and had invited pupils from both foreign and Russian schools in Moscow as special guests of the IKI. Those who didn't speak

Russian adjusted their headsets for the simultaneous translation of the commentary broadcast from loudspeakers.

The audience learned that Vega 2 was into Halley's misty coma, closer to the nucleus than Vega 1, only 4800 miles away from the nucleus at its closest approach. While guests and politicians waited, a statistical chart was displayed on the screens, a listing of Vega's weight, size, and experiments. Everyone, slightly bored, waited for the first pictures of the comet Vega 2 would send down.

The pictures finally arrived, but were just uninteresting bands of purples and blue, no flower-petal center, nothing that to the audience looked remotely like a comet nucleus. Sagdeyev was afraid that his special show for his politicians and schoolchildren might have fizzled.

Downstairs in the operations room, controllers discovered that the platform on which the cameras were attached had shifted. The camera was taking shots too far away from the nucleus.

The cameras were the responsibility of Karoly Szego of the Central Research Institute for Physics, Budapest. Szego transmitted a command to Vega for the camera to increase its field of view and to send back whole frames instead of part of the frame, as it had originally been programmed. This would reduce the total number of images, but the larger field should include the nucleus area and provide something for the VIPs and others upstairs to enjoy.

Vega, however, was 107 million miles away from Moscow. The radio command took 10 minutes to arrive at the craft, 10 minutes more to reply, 20 minutes before the bands on the screen changed into beautiful pictures that to the audience looked comet-like.

The red petal was there, and the fans shooting out from it were yellow, green, and blue. When Vega had approached to 5000 miles from the nucleus, the peanut definitely took shape on the screens, almost as if Halley had two nuclei, two red circles that were connected. One circle was much larger than the other, and waves extending downward from the larger indicated an exploding jet. Generally the DUCMA reported fewer dust impacts with Vega 2 than with its sister craft, and because of this and other evidence, scientists following the dynamics of the flight believed Comet Halley had turned a revolution and a half in the three days between the passages of the two spacecraft. Yet there was still enough exploding dust to knock out 80 percent of the solar panels on Vega 2 as well as three experiments: two spectrometers and a particle analyzer, different instruments from those that had failed on Vega 1.

The Vegas discovered that the temperature of the nucleus surface was between 100 and 250 degrees Fahrenheit in the sunlight. The crust was protecting the ice beneath, which remained at −300 or −400 degrees. Since the crust reflected only 4 or 5 percent of the sunlight that struck it, the investigators knew it had to be really black and that its blackness accounted for some of its massive heat absorption.

Although the two spacecraft were 5000 miles away from the nucleus, their cameras transmitted outlines of two different sides of the revolving nucleus. Afterward, scientists programmed some of their best shots and generated a three-dimensional composite, an object that scaled out to be 9½ miles long, almost 5 miles wide at its widest, and 4½ miles high, give or take a half-mile on each dimension. It was at least a 100 billion-ton peanut, or potato, even an avocado. At the same time, astrophysicists figured, the nucleus was revolving in a 53-hour cycle, reinforcing the data received by Suisei and theorized for Spartan-Halley. There was no evidence for a seven-day rotation, the Russians said.

In a press conference after the flyby, someone asked Sagdeyev why his agency had flown two identical Vegas. He pointed out the instruments that had failed on the first spacecraft and didn't on the second.

"My friends call it Russian roulette," he added and laughed. He was relaxed. His spectacular had been a success. He was still Comrade Cosmos.

During the two days following Vega 2's closest approach, Sakigaki joined the spacecraft flotilla—though at a distance of 4 million miles. Here, its instruments monitored changes in the velocity of the solar wind as well as shifts in the direction of the magnetic fields in space.

Since the seventeenth century, the time of Edmond Halley, astronomers had noticed that every once in a while a comet loses its ion tail but then gains another. It happened to all comets: Halley during the 1910 visit, and Kohoutek in 1974, just to name two. There had been theoretical explanations, and John Brandt, who had been comet scientist on ICE, was at Goddard testing theories of how Halley lost its tail and grew another.

The theory concerned the magnetic field of the solar wind. The field moves in one direction through space, but as the Sun slowly rotates, the magnetic field of the solar wind shifts and establishes a new field, which now goes in the opposite direction. The boundary between the two fields is distinct. When a comet crosses from field 1 to field 2, the ion tail is sheared off and vanishes. But in field 2, a new ion tail is formed. Brandt had recorded that Comet Halley had lost its tail 26 times on its 1980s trip, and on 22 of those occasions the comet had crossed a boundary between magnetic fields.

On March 11 and 12, Sakigake measured such a boundary. On the day Halley crossed it, the comet should have lost its ion tail. But it didn't. Comet Halley didn't lose its tail until March 21.

While Sakigake was crossing Halley's path, the European Space Agency Operations Center at Darmstadt now had five days to guide Giotto into the coma and as close as they could to the nucleus of Halley. But where *was* the nucleus, exactly?

Giotto's exact path was the result of compromise. The principal investigators for the spectrometer and magnetometer experiments on Giotto had said, "As close as possible even if Giotto doesn't survive." The principal investigators for the plasma, dust, and magnetic energy experiments said, "Close enough for good data, but let's survive." Uwe Keller said, "Take the camera between 600 and 300 miles."

The ESA Space Science Department decided to fly the spacecraft about 360 miles off the nucleus. Too close and Giotto might be destroyed, even crash into the nucleus. Too far away and the data and the pictures might be worthless. In addition, there was the aiming and control problem.

Once Giotto was on its flight and inside the bow wave, there was no way to change its path. Even if the controllers in Darmstadt saw on their instruments that Giotto was going to bash into 100 billion tons of a black, crusted, dirty snowball, it would be 8 minutes before the signal to correct that trajectory reached the spacecraft. If Giotto veered off target, Comet Halley would have some fine electrical instruments scattered in pieces over its surface.

To avoid this scenario, the flight dynamic teams of all the other Halley spacecraft cooperated on a combined operation: NASA's deep space network, ISAS at Tokyo, IHW at Pasadena, and the Vegas at Moscow.

These were joined by the International Ultraviolet Explorer (IUE), a tiny satellite, a project of ESA and NASA together, launched in 1978 with an estimated life span of three years. It is still flying and transmitting after 11 years, and along with other targets had recorded the ultraviolet spectrum of 25 comets. After April 1985, it monitored Comet Halley especially during the Vega encounters, and it was also to monitor the Giotto encounter. Measuring Halley's gas and dust production, IUE got an accurate fix simultaneously on the position of the Vegas and of Halley's coma.

The flight dynamic teams of these various projects joined together and developed the "pathfinder concept." Steadily for three days they worked on their computers. They figured exactly where the Vegas were on March 6 and 9, where Yeomans at IHW said the comet was on those dates, and where the Vegas had reported that the nucleus was. With this and other information the computers projected where the nucleus would be on March 14, and telefaxed this information directly to ESOC, Darmstadt.

As a result of this international cooperation, the controllers at Darmstadt were able to reduce their target area for Giotto from hundreds of miles to a bull's-eye on Halley's coma only 24 miles in diameter. Thus they could more accurately aim Giotto to fly exactly 360 miles away from the nucleus.

In the evening of March 13, at about 8:45, a radio signal left Darmstadt for Giotto, now 86.4 million miles above West Germany and some 600,000 miles away from the nucleus of Comet Halley. The signal didn't arrive for 8 minutes. It went from Darmstadt to a telecommunications satellite, from the satellite to the Carnavon receiving station on the Indian Ocean in Western Australia, across the Australian continent to Parkes, New South Wales, and from Parkes to the dish on Giotto, provided that dish was pointed toward the Earth. On March 13, the dish was so lined up. The camera and the other instruments received the signal and instruments switched on. Sixteen minutes later the signal from the spacecraft transmitting along the same circuitous route, was received at Darmstadt and was broken into 10 streams, one for each instrument.

The operation center at Darmstadt is tiny compared with the gigantic facilities at Moscow. It has about one-tenth of the personnel and equipment as the IKI. Two rows of computer consoles at Darmstadt monitored the Giotto payload, and the atmosphere at Darmstadt was that of a family reunion at Christmas.

Frankfurt, West Germany's New York, is only 15 miles from Darmstadt, and already TV trucks had gathered around the Center and 2000 visitors were milling around the grounds, fortunately not in the operations center itself. A Eurovision broadcast of Giotto's very close encounter would soon go to 50 nations to be seen eventually by more than a billion viewers, more than any other program ever seen.

A map of Giotto's trip was projected on a small screen, the bow wave, coma, nucleus, and a line drawing of Giotto on the path it would take. The map showed Giotto had just penetrated into the turmoil of the coma, which is the comet's atmosphere, 590,000 miles from its closest approach.

Soon the instrument signals arrived on the consoles. Scientists immediately crowded around the screens. About three people could watch a single CRT, but always a fourth or a fifth scientist joined behind the others, moving around trying to read the math on the screen.

Giotto was soaring into the comet's atmosphere, still 1 hour and 10 minutes from its closet approach to the nucleus, 174,000 miles out. J. Anthony M. McDonnell, from the University of Kent, England, principal investigator for the dust detector, DIDSY, watched his console as the sensors on the outer shield of Giotto were peppered with dust particles, tiny ones, but many of them.

An hour later, the first images from the camera arrived from the spacecraft on a small wall screen with a sign above it, "HMC," for Halley multicolor camera. At this distance, the camera gave few details, but it transmitted the position of the spacecraft relative to the nucleus. With the spacecraft 3 hours away from its closest approach, the team readjusted its position according to the information from pathfinder concept.

The camera, fixed on its platform, revolved with the spacecraft once every 4 seconds, photographing a circle across the distant coma. One of the computers directed the camera to focus on the brightest spot ahead. Alan Delamere and the other Giotto programmers, like the Russian programmers, hadn't known 3 years before that the nucleus was not the brightest spot. Still the field of view of the Giotto camera was wide enough that it soon caught the nucleus as a dark shadow.

The camera focused only a few milliseconds during each spin, but the CCD's permitted a ten millionth of a second bit, time for a picture without smearing. Another computer told the camera where to focus next, and the process was repeated eventually more than 2000 times. If the three microprocessors worked properly, picture taking was automatic. Keller and the team could just sit back with their hands in their pockets watching the screen.

The picture developed in strips, one strip building on top of the other every four seconds as Giotto rotated. The first pictures on the screen were black, relieved by a brilliant spot of white, a floodlight spreading out against black velvet. The camera showed the jets shooting sunward as a single explosion of light.

For the next projections the computers converted the picture into radiant eddies of greens, reds, blues, purples, an abstract canvas of computer art. Numbers transmitted from Giotto were computerized into colors as they had been for the Vegas, fans of yellow, red, and blue, actually dust jets shooting out of the nucleus into the coma.

When the spacecraft was still around 75,000 miles and 30 minutes from the nucleus, it had been hit more than 11,000 times with both small- and medium-size particles. However, none of them had penetrated that first shield. The outer shield prevented any ion build up too. Inside the body of Giotto, the instruments and computers were functioning perfectly.

As Giotto continued through the coma, two of the DIDSY stations were saturated with an inundation of tiny particles. Then the dust activity slackened but it picked up again when Giotto was 42,000 miles and 17 minutes away from the nucleus. The pictures on the screen were unaffected and unfolded one right after the other. As soon as one set of strips filled the area, the picture vanished and another started.

Giotto was moving at about 148,000 miles per hour, and soon the peanut shape of the nucleus emerged on the screen in a blue outline, and a single jet flared off the blue in waves of different light intensity represented by yellow and red. In each set of frames the nucleus was larger, the colors primitive, brilliant, and stark.

At the same time, the dust particles within the coma were getting larger. Just beyond 3 minutes before the closest approach, DIDSY reported the first penetration of the outside shield—eleven hits on the inner Kevlar shield. Some pictures arrived on the screen at Darmstadt without color enhancement. The daylight side of the nucleus, which was turned toward the Sun, glittered so that it was impossible to pick out any details of the nucleus surface. Only along the line between daylight and night, the terminator line, could the audience make out irregularities in Halley's topography.

Wolfgang Schmidt, one of the scientists from the Max Planck Institute and a member of the camera team, doubled as the English-language TV broadcaster. He explained that the most prominent of the irregularities on Halley's nucleus was a "mountain," 1300 feet high, probably Halley's Everest. Other irregularities could be foothills. Schmidt pointed out craters as well, but these craters had not been formed by impacts, but rather were the sites of previous explosions from *beneath* the crust, like erupting volcanoes of dust and gas.

The color returned to the screen. The bright dust jets surrounded a white center with shimmers of blues and purples with the shape and look of the interior of an oyster shell. At 2400 miles, a minute away from its closest approach to the nucleus, the camera focused as if it were only 330 feet away.

However, as the Giotto was soaring into the coma, the spacecraft ran into a new cascade of dust. More than 100 particles struck the satellite shields. Forty-four seconds away from the nucleus, only 1800 miles out, DIDSY recorded being struck by the largest piece of dust of the entire journey, a vertible hunk weighing one twelve thousandth of an ounce. Giotto's velocity, and that of the dust particle, gave the encounter the impact of an M-16 bullet hitting its target. Twenty-two seconds later, small dust particles impacted on the metal mirror of the camera and made the TV picture fuzzy.

Suddenly, when Giotto was only 700 miles away from its closest approach, the TV screens blanked out. Giotto had been struck by another large piece of dust. The spacecraft began to wobble like a top losing its speed, weaving back and forth, a wobbling called nutation. The wobble moved the transmission dish out of alignment with Earth and specifically out of communication with Parkes station, Australia.

Another piece of dust penetrated both shields and damaged one of the camera's computers. The computer returned to the beginning of its program just as if it were starting the mission all over again. The computer directed the camera to search for the comet. The camera searched in vain. It had already passed the comet.

During the wobble Parkes and NASA's deep space network station at Canberra received intermittent, weak signals from the instruments. Fortunately these signals were recorded on tape and eventually were read out.

On board Giotto, a damper began slowly to stabilize the craft. The damper was a backup installed because somebody had thought about the possibility of having to eliminate the very nutation that occurred. In 34 minutes nutation stopped and radio telecommunications came through clear again. Keller, Reitsema, and the team tried to communicate with the camera, but it did not respond to commands. Blindly, it continued as ordered by the microprocessor, searching for the nucleus, now 84,000 miles behind it. In spite of what the microprocessor was attempting to do, Giotto's Halley mission had been completed.

Robot spacecraft had entered the atmosphere of comets and signaled what they had observed to Earth-bound observers. Sailing on, the spacecraft left Comet Halley behind them and continued flights through space.

The spacecraft weren't left on their own to wander aimlessly through the cosmos. ISAS borrowed Bob Farquhar from Goddard and brought him to Tokyo to establish new paths for Sakigake and Suisei. The new track for Sakigake, the pioneer, will return it to the vicinity of Earth January 8, 1992, then whip it around the planet four times. During these trips Sakigake will transmit reports on the Earth's magnetotail.

The four flybys are intended to whip the spacecraft within

6,200 miles of the Comet Honda-Mrkos-Pajdusakova on February 4, 1996, a comet with a short orbit of five years discovered in 1948 by a Japanese and two Czechoslovakian astronomers.

The Goddard flight dynamics team also fixed Suisei's path to bring it back to the vicinty of the Earth, August 20, 1992. The Earth's gravity will target the spacecraft to Farquhar's old friend, Giacobini-Zinner, to observe how it may have changed.

As for Giotto, as it flew away from the vicinity of Halley, its microprocessor was still instructing the camera to locate the nucleus. The controllers at Darmstadt deactivated the camera. Later, they discovered that the camera's baffling tube was missing. The tube had originally extended at a right angle to the camera itself to protect it from dust and sunlight.

Half of the instruments on Giotto were still functioning, and the mission's science committee decided to send it on another comet mission. David Dunham laid out a path that will swing Giotto by the Earth July 2, 1990. Then ESA will have a choice of mission for the spacecraft. It will be able to reach Comet Hartley 2 in August 1991, Comet Grigg-Skjellerup on July 10, 1992, or another comet with a five-year orbit. The camera, still searching for a nucleus, can now find one ahead, as it once did on the Halley mission.

Sadgeyev and the IKI decided that the damage to the two Vegas should also not prevent other missions. Tentative plans are to have the Vegas scan one of the larger asteroids, Adonis, from 3.7 million miles away, measuring the plasma in its environs. If possible, the spacecraft will continue their voyage to approach Jupiter and transmit photo images of the planet.

After its invasion by the space flotilla in March, Comet Halley put on its finest show for observers on Earth, at least for those who were in the best location—which was south of the 20th North Parallel and further south in the Southern Hemisphere.

The area south of the Tropic of Cancer presented a particular problem to the International Halley Watch setting up its world-girdling network. A large percent of the Southern Hemisphere is ocean.

John Brandt with Malcolm Niedner of Goddard and Jurgen Ray of Bamberg, had to establish observatories in remote and far separated areas: Tahiti; in the Indian Ocean, off the east coast of Africa, Reunion Island; off the west coast of South America, Easter Island, known for its huge statues; and Faraday Station in Antarctica, among others.

Each observatory had to be equipped with portable equipment—16-inch reflectors are considered portable—a photography studio capable of developing black and white prints, and clock drives that run backward as compared with those in the Northern Hemisphere.

The astronomer assigned to Reunion Island survived a hurricane and an erupting volcano. On Easter Island the observatory was a box made from the metal panels of a disassembled satellite tracking station. The top of the box opened for viewing then closed when it was not in use. In Anarctica the astronomer arrived a month before his telescope did.

In spite of their troubles, the International Halley Watch achieved the coverage it wanted. For example, from his box on Easter Island, William Liller made spectacular shots every day from March 1 through April 19, a series of 48.

Other astronomers, like King Arthur's knights, ranged widely on their quest. None of them did so on horseback, but rather in trucks and cars, on ships and planes. Their weapons varied also, from pearl-covered opera glasses to 8 by 50 binocular giants, which Sir Lancelot had to hold without swaying to scan the Southern sky. And, of course, every kind of telescope was used: a nineteenth century, long brass refractor with Zeiss optics still sharp; Newtonians and Cassegrains ranging in apertures from 4 to 11 inches. The smaller telescopes had an advantage. They could be carried on the plane and would not be lost somewhere in the Atlanta or Dallas airports when the astronomers deplaned at Mexico City, Quito (Ecuador), Papeete (Tahiti), or Sidney (Australia).

Wherever they arrived, the questing knights found the Holy Grail was there, scintillating in the morning twilight in March and, after April 15, glowing against the velvet night sky, its tail lengthening on each night.

In the towns of Ecuador, in the mountains of Chile, in Mexico, Uruguay, and Argentina, ordinary people were excited by Comet Halley. These were the early risers, hard-working people, such as a farmer on his way to market with a cart of chickens. He was pulling the shafts of the cart himself. He looked up to see the comet above him.

A mother in a heavy blue skirt was driving a group of pigs down the road with a long willow switch. Her young son was helping her. He looked up. He jerked on her blouse sleeve. "*Mire, mire,*" he called. He pointed with his switch. She looked up from the pigs and nodded violently.

There was a coffin maker in Colima, south of Guadalajara. His shop was only a roof on stilts piled with half a dozen unpainted coffins. Under the weak light of a hanging lantern, he was shaping a new lid with a heavy, wooden plane.

Four gringo astronomers from Reno, Nevada, staying at the hotel in Colima, walked past his shop long before sunrise to set up in a field not far away. They were harnessed with telescopes, cameras, and bags of accessories. One of them had an Astrocan, a red telescope with a bulbous base, slung across his back, and he wore a Halley T-shirt, blue with a brilliant, stylized comet.

The coffin maker recognized it and waved and cheered. After they had set up and were photographing the brilliant comet above them, the coffin maker arrived with a jug of tequila mixed with pomegranate wine. He insisted on partying. Soon the Nevadans were brighter than the Comet.

On Mt. Thorodin, Colorado, Gary Emerson and his wife, Paula, loaded the old 9-inch, dumpster-retrieved telescope-camera into their station wagon. They drove down to Big Bend National Park in Texas and camped in the mountains just south of Terlingua. Each morning Emerson, his wife, and some other astronomers got up while the sky was still dark and clear to set up their telescopes and cameras.

About 4 A.M. they watched the tail of the comet rise from

behind the mountains. Suddenly the coma was in the sky, and for 45 minutes the comet dazzled, the most brilliant cosmic apparition. The astronomers were silent, worshipful. In the sky above them were the 4 billion years since the formation of the solar system, the 76 years since 1910, and the 76 years until 2061.

Comet Halley has been a legend, a spirit remembered, a hope in dreams. That moment when its radiance hovered over the black outlines of the Chisos Peaks of Texas, Comet Halley was the brush with eternity; its light pierced the cloud of the unknowing.

There was a woman living near the Emerson's camp who had glaucoma. She didn't get up the first few mornings because she just assumed she couldn't see the comet. Finally Paula Emerson talked her into getting up with them. Halley was so bright that she could see it with her naked eye. She jumped up and down. Every morning from then on she got up and went out with the Emersons, comparing the comet's brilliance with what she'd seen the day before.

The last night the Emersons were in Terlingua, they took the woman out to dinner. Terlingua is a small town, and the restaurant isn't big either, a bar on one side of the room and maybe a half dozen tables between the bar and the door. During the meal they could hear the conversations all around them. Everyone was talking about Comet Halley. It seemed they were all getting up in the early morning and looking at it. It was the most exciting thing that had happened in Terlingua for a long, long time.

19

Deciphering the Cosmic Rosetta Stone

The appearance of Comet Halley, predicted and engrossing, was the spectacular celestial event of 1986 not only to the casual observer like the little boy driving his pigs but also to the serious amateur who contributed to the library of the International Halley Watch at Pasadena, California.

Steve Edberg, who had organized the amateurs for the IHW, received 337 separate reports from amateurs in 54 different countries, in alphabetical order from Argentina to Zimbabwe, and in between China, Kuwait, Mauritius, India, and Ruwanda. There were 122 reports from Italy, 92 from England, 85 from East Germany. Most reports, 442 altogether, were from the United States. The library archive contains hourly sightings of the comet from the middle of August 1985, when Comet Halley moved away from the glare of the Sun, until October 1986, when it appeared as a ball of fuzz at a magnitude of + 13. IHW plans to publish 30 volumes of data, 1000 pages a volume, issued as compact disks, microfilms, and printed summaries. The library of the IHW will provide the most complete report of a comet that has ever been assembled.

The robot spacecraft that visited the comet had the advantage of not having to sight through the Earth's blanket of atmosphere, and for the first time the heart of the comet had been revealed to humankind. The nucleus of Comet Halley was discovered to be the shape of a peanut or a potato, and bigger than most astronomers expected. Fred Whipple called it relatively the size of Manhattan Island, about 10 miles from where Halley's Harlem River should be to its theoretical Battery Park; about 5 miles from an imaginary Hudson River to an imaginary East River. Then there's an additional dimension for the flying Manhattan, down through the body of the nucleus, about 4¾ miles.

Harold Reitsema and Alan Delamere, two of the engineers who designed the Halley multicolor camera for Giotto, wanted to create the very first picture of a nucleus as it might appear if it could be seen without the gases and dust of the coma interfering. With the help of their computer at Ball Aerospace, they synthesized 60 of the best pictures the camera had transmitted while Giotto was cruising between 12,500 and 2500 miles from the nucleus (during that trip, the camera had recorded details as if it had been only ½ mile to 330 feet away).

Light reflected off the dust of the nucleus surface silhouetted an outline of the nucleus against the black of space. The nucleus does seem to have the general shape of a peanut, or a potato (perhaps offering even a wider choice of vegetables). The bright jets of dust are exploding as the Sun strikes the surface facing it. The line between Halley's night and day can be observed. Surrounded by the dark, Halley's 1300-foot "mountain," a bright oblong of varying intensity, has caught the sunlight before its surrounding area. Finally we have been provided with a look into the heart of a comet, the tiny body which produces such a magnificent show, as Fred Whipple had pointed out at the Goddard conference in 1977.

All the spacecraft from Pioneer Venus to Giotto recorded

evidence of the Comet's rotation in space. Scientists brought their differences of opinion about the length of Halley's day to two conferences, the first held at Heidelberg University in October 1986, the second at Paris in November. Jack Lissauer, from the University of California at Santa Barbara, bought a melon at the Heidelberg market in the approximate shape of the Comet. Lissauer tossed his melon up and down, and everyone watched how it moved. It rotated as they expected, with the same kind of spin quarterbacks give to a football.

But the melon moved in another way, too. The whole melon also wobbled, a nutation, the same motion that Giotto experienced when hit by dust. The two movements could be explained according to the formulas of the eighteenth-century mathematical genius Leonard Euler. According to these formulae, Halley rotated every 54 hours, its day-night cycle. Its nutation, however, had a period of 7.4 days, the period Pioneer Venus had discovered on its data. Not everyone at either Heidelberg or Paris was convinced Halley moved like a melon, however, and the debate continues even today.

Like the comestibles to which it has been compared, Halley is covered with a skin, a shell, a crust. Whipple said the crust was blacker than coal dust, like black velvet, the blackest object in the solar system. The Vegas and Giotto reported its temperature on the sunlit side between 100 to 250 degrees Fahrenheit. On this side the crust sometimes cracked open, and through these cracks exploded the jets Vega and Giotto encountered. The jets were 80 percent water vapor which at their most violent released 30 tons of liquid a second. The jets extended 100 miles into space and worked much like the thrusters fired to guide spacecraft. They altered Halley's path and have caused some of the difficulties encountered in trying to exactly predict the comet's orbit and perihelion.

Simpson's DUCMA on the Vega spacecraft reported that the jets carried the small particles of dust into the coma and tail: the larger particles, they surmised, fell back and helped form the crust. Altogether the scientists figured that Halley lost close to 165 million tons of its mass during this trip past the Earth and Sun. They estimate that the Comet still has enough mass to endure 253,308 years more and so will appear for 3300 more visits.

Has Whipple's image of a comet as a "dirty snowball" stood up to the new evidence? Whipple wrote in *Sky and Telescope*, March 1987, that instead of being a dirty snowball, Halley "more resembles a dirty snow*drift*," its density very much like newly fallen snow, one-third to nine-tenths empty space.

Comet Halley as a "dirty snowdrift" has implications for our view of the Comet as a Rosetta Stone, the key to deciphering our cosmic past. To account for its being so loosely packed, it must have been formed within the debris nebula that collapsed to form the Sun and the planets when temperatures were near −500° Fahrenheit. This temperature is close to 100 degrees of the temperature of the inner nucleus, beneath the hot, black surface.

The nucleus was molded into a potato or peanut shape by low-speed collisions while it was circulating in the vicinity of the outer planets and was still part of the debris nebula. The inner planets consist mostly of rock. Among the outer planets, Saturn, Uranus, and Neptune, ice is the basic constituent. The rings of Saturn and Uranus are ice too.

Dust particle analyzers and spectrometers on both the Vegas and on Giotto identified the dust and gas as made up of carbon, hydrogen, oxygen, and nitrogen, (the four elements with the acronym CHON), carbon-based material, formed into various molecules. These four CHON elements are also important constituents of the satellites of the giant planets, Jupiter and Saturn, and of the planets Uranus and Neptune. The CHON dust particles were apparently the source of the cyanogen that caused the panic during Halley's passage in 1910. Other dust particles contain Earthly elements, sulfur, silicon, and magnesium, found in certain kinds of rocks here on Earth.

An ion spectrometer on Giotto made the important identification of ammonia and methane in the water vapor of the

Comet's gases. The molecules that form these two substances had not been identified in the gases of Kohoutek in 1973. Ammonia and methane, when subjected to any form of energy—an electrical charge, heat, even shock waves, will synthesize into amino acids. The presence of ammonia and methane in the gases of Halley establishes the possibility that at least some comets carry amino acids, and thus may have brought to Earth the basic components of primitive proteins out of which life can form. As we recall, certain amino acids were discovered in the residue left by collisions between Earth and a space object 150 million years ago, collisions which, it has been theorized, ended the existence of dinosaurs.

An instrument on Giotto, the positive ion cluster composition analyzer (PICCA), transmitted data that further confirms Comet Halley as a Rosetta Stone. The PICCA, built cooperatively by the University of California at Berkeley and the Max Planck Institute, Lindau, West Germany, detected molecules in the coma of Halley that proved to be polyoxymethylene (POM), also known as polymerized formaldehyde. Comet Halley could only have acquired formaldehyde when it was still part of the debris nebula and before it was ejected out of the newly forming solar system into Oort's Cloud. (Incidentally, formaldehyde was one of the first plastics and was made into those neck-chafing collars worn by subjects stiffly posed in early photographs.)

Scientists are still analyzing the data gathered by the spacecraft and by ground-based telescopes and will continue to do so for many years to come. These analyses expand our knowledge of the composition of comets, but at the same time they pose new questions for future investigations.

Now we have advanced astronomical instruments ready to continue detailed investigations into our cosmic origins. On September 29, 1988, with the launch of *Discovery*, the shuttles started flying again. Probably about spring 1990, the Hubble space telescope is scheduled to lift into orbit 350 miles above the Earth. Financed jointly by ESA and NASA, the telescope was named after Edwin Hubble (1889–1953), the astronomer who first formulated the concept of an expanding universe.

When the Hubble space telescope is in orbit, astronomers will have an eye in the clear vacuum of space above the atmosphere. The telescope is a cylinder 43 feet long and 14 feet in diameter. Its main mirror is 94 inches, which puts it in a class with some of the largest telescopes on Earth. The Hubble's power will be supplied by solar arrays, the responsibility of ESA. On its CCD's it will not only be able to record visual light, but its spectrometers will be able to pick up the ultraviolet and infrared bands. One of its microprocessors has been programmed with a global star catalog. All these and other support instruments will be carefully maintained by a staff of 320 and can be repaired or replaced by astronauts operating from one of the shuttles.

The telescope will be operated from the Space Telescope Science Institute at Johns Hopkins University, Baltimore. Already astronomers waiting to use the telescope are on lists that extend into the last years of the 1990s. The European Southern Observatory at Munich will coordinate, preserve, and make available the data the telescope collects for astronomers who missed the waiting list.

Followed in space by the eye of the Hubble telescope, Comet Halley can never vanish. Astronomers will be able to follow it to its aphelion, right out to that point beyond Neptune where in 2024 its velocity weakens so it turns the corner and is pulled back again toward the Sun and Earth.

The Infrared Astronomical Telescope, which NASA launched in 1983, detected a debris nebula around the star Vega, which is only 27 light years away from Earth. Atop a mountain in Chile a 100-inch telescope detected a disk around a star 50 light years away in the southern constellation, Pictor, near the Large Magellanic Cloud. Using the Hubble telescope, astronomers will be able to analyze these nebulous disks and undoubtedly discover others. Are these disks solar systems in the process of formation 4½ billion years after our own experience? Ten different stars have given evidence that they may have planets. The Hubble telescope will confirm the presence of other planets, if we really do have neighbors circling nearby suns.

While the Hubble space telescope will be looking out, the Comet Rendezvous Asteroid Flyby (CRAF) will be flying out to continue the search for our cosmic past that the Halley spacecraft had begun. To be built jointly by NASA and firms in West Germany, CRAF will carry 19 separate experiments organized by project scientist Marcia Neugebauer of JPL. Presently scheduled for launch on a shuttle in 1995, CRAF will be the first mission to be flown on the Mariner Mark II, one of two standardized spacecraft. The Mark II is a development of the Spartan system. It supplies the power propulsion and guidance systems, and Neugebauer will integrate the CRAF instruments with the Mark II carrier as Fred Wilshusen had to integrate the spectrometers and cameras of Spartan-Halley into the Spartan carrier.

The CRAF project has experienced innumerable delays. With each delay its comet target has to be changed for one available according to the launch date. Earlier, CRAF was scheduled to be launched in 1993 on a Titan-Centaur rocket and, after a gravity assist from Venus, then the Earth, the spacecraft would pass through the asteroid belt and rendezvous with Tempel 2 in October 1996.

With the new launch date, CRAF is still scheduled to fly through the asteroid belt and scan the asteroid Iris along the way, then continue on to rendezvous with a comet.

Iris, a much larger asteroid than Adonis, the target chosen for the Vegas, is about 390 miles in circumference as compared with about a third of a mile for Adonis. CRAF will photograph Iris and catalog surface minerals seen in infrared light.

The asteroid missions are a natural extension of the Halley missions. Anything we can learn about the asteroids will supplement what we have learned about the origin of the solar system from Comet Halley.

There are 3700 numbered asteroids, but their actual number is much more than that: there are perhaps even 100,000 of them, some so tiny as to be invisible to us. Asteroids are actually minor planets, orbiting in a "belt" between Mars and Jupiter and observed from the Earth to be orbiting in a counterclockwise direction like the planets but opposite to the direction of Comet Halley. They vary tremendously in size between those so small as to be invisible from Earth to the largest, Ceres, 2200 miles in circumference. Originally astronomers thought that the asteroids were the remnants of a planet that self-destructed, like Superman's Krypton. In spite of their numbers, the asteroids don't constitute enough mass for a planet. A more recent theory posits that asteroids are left over from the original debris nebula, matter that somehow never coalesced into a planet. The comet storms that impacted primitive Earth and possibly left oceans and atmosphere as a result, first smashed through the asteroid belt. It is also theorized that the asteroid belt is the immediate source of the meteorites that have collided with the Earth. The CRAF mission will add to what we know about the asteroids and help determine just exactly how they fit into the origin of the solar system.

CRAF is now scheduled to rendezvous with Comet Kopff, a comet with about a six-year orbit. First discovered in 1906, it has been recovered regularly since 1919. Arriving at Comet Kopff in 1998, CRAF will match its trajectory exactly to the orbit of the comet and will be able to observe Kopff as if it were stationary. Using both visual camera and infrared spectrometer the spacecraft will map the entire surface of the comet, revealing details that are only 3 feet across. Twelve experiments will record the physical properties of the crust: minerals and temperatures. Hovering nearby, CRAF will fire a probe containing five instruments 3 feet through the crust and into the interior of the comet. A small sample will enter the probe, be sealed, and then brought back into the spacecraft for various analyses, the results telemetered to mission control on Earth.

Neugebauer and the principal investigators of the experiments expect to resolve some of the questions that the flybys of Comet Halley raised. Comet Halley is covered with a black, heat-absorbing crust, but is it a light layer of dust which has fallen there, not driven into the coma and dust tail by the forces of the Sun? Or a solid from which volatile material has boiled away?

Protected by the crust, the interior is a pristine remnant of the solar system. How tightly compacted is the interior? Was Whipple correct when he called the stuff of a comet a "snowdrift?"

The spectrometers on all the spacecraft that surveyed the coma of Comet Halley registered various ions and molecules. As we know, however, everything recorded had been affected by the magnetic field of the solar wind. The spectrometer in the tip of CRAF's probe, having penetrated the nucleus of Comet Kopff, will be able to detect the "parents" of these "daughter" substances before the parent had been ionized by the solar wind. If, as Delsemme suggested at the Goddard Conference in 1977, in the nucleus of a comet are discovered some of the gases rare in our atmosphere, xenon, for example, it would strengthen the theory that comets did bring the atmosphere to primitive Earth.

As we know, one of the spectrometers carried by Giotto discovered ammonia and methane in the water vapor of Halley's coma. These gases indicated the possibility of amino acids being present in the nucleus. There are many different kinds of amino acids. If CRAF discovers the presence of the amino acids essential in protein molecules in the nucleus of Comet Kopff, that discovery would establish the possibility that comets did bring to Earth the basic ingredients of life's primordial soup. Such a discovery would truly make the CRAF mission as memorable as the 5-year cruise of the *Beagle* (1831–1836) and realize Delsemme's hope for a mission rivaling that of Darwin's.

In 1981 NASA did not have time to construct the Halley Earth Return mission that Farquhar proposed. From July 15 to 17, 1986, scientists and engineers from Europe and the United States gathered at Canterbury, England, for a Comet Nucleus Sample Return Mission workshop. The conferees, including Farquhar, pooled ideas for targets, for spacecraft, and the advance in technologies that will be required. Between now and the end of the twentieth century, only short-period comets will be available, some of these over five astronomical units away, a trip 10 times the distance the Halley spacecraft had to travel.

None of the spacecraft designs the scientists offered were simple, but some were more complicated than others. All the spacecraft were unmanned. Some, like CRAF, were designed as hovering above the comet and firing a penetrator, then retrieving it. A moving craft would allow more than one sample, because the inner nucleus of a revolving comet is probably not uniform.

Other designs called for landers that descend at a prescibed rate then set down on cone-studded legs, which are preheated and vented to allow any gases to escape. However, drills of any type, coring drills, coring tubes, must operate at very low temperatures without heating the sample. Then the samples must be put in a canister and their temperature maintained over the long trip back to Earth. Such a trip might take four or five years, and the spacecraft carrying the cannister could experience great changes in outside pressures and temperatures.

As complicated as the designs were, they are not science fiction. They are all within our engineering capabilities at the present time. Actually, the Galileo spacecraft, scheduled to be launched October 12, 1989, is more advanced than any of these. It will carry more sophisticated experiments to Jupiter and Jupiter's moons.

How far will we have advanced in 2061, when Comet Halley returns? Seventy-six years is long enough to be a unique experience in most people's lifetime, and it is still short enough so the comet is never forgotten. In 1985 there were a few special people who could recall Comet Halley's visit in 1910–1911. In 2061 there will be other special people who will remember it from 1985–1986. If Halley's time span had been 150 or 200 years, this continuity would not have been possible.

Each of Halley's 76-year cycles forces us to evaluate our advancement in science, engineering, and technology. In 2061 the second, possibly the third, generation of nucleus samples will have been returned from other comets. These samples would not have to be delivered to an Earth-bound laboratory, but would be able to be returned to a laboratory in a huge international space station on orbit 200 miles in space.

Twenty-one years after Skylab's demise, a shuttle is scheduled to begin transporting the first elements of American's second space station, Freedom, developed from the first space station, Skylab. Beginning about in 1995, sections of prefabricated structure will be put together 200 miles above the Earth. Here astronauts will assemble a structural frame 300 feet square. The complex scaffolding will eventually support two 150-foot solar panels that will provide power for interconnected space laboratories and living quarters. The laboratories and the astronauts' home will be twice the size of Skylab's Saturn IVB module, but the astronauts on Freedom will have the same daily needs as those on Skylab, and NASA can use the experience they gained in 1978, supply good food, comfortable living quarters, and a much improved communications system, which will make shouting unnecessary. NASA has so far refused to commit itself on serving the beer and wine that Pete Conrad, commander of the first Skylab crew, argued for so eloquently.

However, Freedom will be an international space station. The American module will be joined by others from Canada, Japan and the European Space Agency. Hermes, the ESA shuttle, will be able to dock at Freedom and also at Mir, the Soviet space station, which by the end of the century will be greatly expanded.

The crews on Freedom will be able to continue the kind of

studies begun on Skylab, studies of the Earth, of the Sun, and of comets. They will be able to conduct technically complicated experiments and engage in the manufacture of materials possible only in the zero gravity of space.

In addition, Freedom will be able to operate as a space port, an assembly and transfer point for manned and unmanned voyages, as nations explore further into space. In 1987, Sally Ride, the first American woman astronaut to fly in space, resigned and published her book, *Leadership and America's Future in Space*. In it she outlined a realistic, four-step space program which culminates in a manned Mission to Mars.

The recent loss of the two Soviet Phobos craft sent to Mars demonstrates the severe challenge of that trip: 300 days, 61 million miles. Although Roald Sagdeyev has resigned as director the Soviet Space Research Institute, he is still serving as a member of the Soviet Academy of Sciences. His strong belief in the cooperation of those nations involved in space exploration is continuing. To aid IKI in its planning for its Phobos mission, NASA supplied it with our data on Mars from our Mariner and Viking missions. For our Magellan mission, IKI supplied NASA with data it had gathered during the Venera 15 and 16 missions to Venus in 1983. This cooperation between IKI and NASA will continue with our Mars Observer, a spacecraft scheduled to begin its flight toward that planet in 1992, taking advantage of the lessons both agencies learned from the Phobos experience.

The Phobos craft carried, among others, 15 American-sponsored experiments, and Larry Esposito, astronomer at the University of Colorado at Boulder, reported that most of his experiment, monitoring dust particles in the Martian atmosphere, had been completed before the loss of communication. Before Phobos I was misdirected and sailed out of its planned orbit, it had already telemetered information it had gathered on the x-ray emissions of the Sun.

By 2061 our accumulated experience should have placed

scientists from many nations on Mars. Here, in addition to other experiments, the scientists will be able to monitor the coma and tails of Comet Halley just as they start to develop.

The search for our cosmic origins will not end on Mars. It can never end. Each scientific advance leads to the next. Comets have been analyzed by the eyes of Aristotle, by the quadrants of Tycho Brahe, by ever larger telescopes, by ever more sensitive spectroscopes, from spacecraft, from space stations.

The scientists and engineers of yesterday were preparing the way for the scientists and engineers of today. The scientists and engineers of today are preparing the way for those of tomorrow. They all are helping to fulfill the hope Robert F. Kennedy expressed in 1964:

> The world should be a better place when we turn it over to the next generation than when we inherited it from the last generation.

Index